The Exchange Format Handbook

A DEF, LEF, PDEF, SDF, SPEF & VCD Primer

Other books by the same author:

- **In Chinese**: *A SystemC Primer*, Tsinghua University Press, China, ISBN 7-302-08418-1, 2005.
- *A Verilog HDL Primer, Third Edition*, Star Galaxy Publishing, Allentown, PA, 2005, ISBN 0-9650391-6-1.
- **In Chinese**: *Verilog HDL Synthesis, A Practical Primer*, Tsinghua University Press, China, ISBN 7-302-07714-2, 2004.
- *A SystemC Primer, Second Edition*, Star Galaxy Publishing, Allentown, PA, 2004, ISBN 0-9650391-2-9 (First edition, 2002, ISBN 0-9650391-8-8).
- **English edition for Indian subcontinent**: *A VHDL Synthesis Primer, Second Edition*, BS Publications, ISBN 81-7800-014-8, 2001.
- **English edition for Indian subcontinent**: *A Verilog HDL Primer, Second Edition*, BS Publications, ISBN 81-7800-012-1, 2001.
- **English edition for Indian subcontinent**: *Verilog HDL Synthesis, A Practical Primer*, BS Publications, ISBN 81-7800-011-3, 2001.
- **In Japanese**: *Verilog HDL Synthesis, A Practical Primer*, CQ Publishing (http://www.cqpub.co.jp), Japan, ISBN 4-7898-3354-2, 2001.
- **In Chinese**: *A Verilog HDL Primer, Second Edition*, China Machine Press (http://www.hzbook.com), ISBN 7-111-07890-X, 2000.
- *A VHDL Primer, Third Edition*, Prentice Hall, Englewood Cliffs, NJ, 1999, ISBN 0-13-096575-8.
- *A Verilog HDL Primer, Second Edition*, Star Galaxy Publishing, Allentown, PA, 1999, ISBN 0-9650391-7-X.
- *Verilog HDL Synthesis, A Practical Primer*, Star Galaxy Publishing, Allentown, PA, 1998, ISBN 0-9650391-5-3.
- *A VHDL Synthesis Primer, Second Edition*, Star Galaxy Publishing, Allentown, PA, 1998, ISBN 0-9650391-9-6 (First Edition, 1996, ISBN 0-9650391-0-2).
- *A Verilog HDL Primer*, Star Galaxy Press, Allentown, PA, 1997, ISBN 0-9656277-4-8.
- **In German**: *Die VHDL-Syntax* (Translation of *A Guide to VHDL Syntax*), Prentice Hall Verlag GmbH, 1996, ISBN 3-8272-9528-9.
- *A VHDL Primer: Revised Edition*, Prentice Hall, Englewood Cliffs, NJ, 1995, ISBN 0-13-181447-8.
- *A Guide to VHDL Syntax*, Prentice Hall, Englewood Cliffs, NJ, 1995, ISBN 0-13-324351-6.
- *VHDL Features and Applications: Study Guide*, IEEE, 1995, Order No. HL5712.
- **In Japanese**: *A VHDL Primer*, CQ Publishing, Japan, ISBN 4-7898-3286-4, 1995.
- *A VHDL Primer*, Prentice Hall, Englewood Cliffs, NJ, 1992, ISBN 0-13-952987-X.

The Exchange Format Handbook

A DEF, LEF, PDEF, SDF, SPEF & VCD Primer

J. BHASKER

eSilicon Corporation

Star Galaxy Publishing

Published by:

Star Galaxy Publishing, 1058 Treeline Drive, Allentown, PA 18103

Phone: 888-727-7296, Fax: 610-391-7296, http://www.stargalaxypub.com

---

WARNING - DISCLAIMER

The author and publisher have used their best efforts in preparing this book and the examples contained in it. They make no representation, however, that the examples are error-free or are suitable for every application to which a reader may attempt to apply them. The author and the publisher make no warranty of any kind, expressed or implied, with regard to these examples, documentation or theory contained in this book, all of which is provided "as is". The author and the publisher shall not be liable for any direct or indirect damages arising from any use, direct or indirect, of the examples provided in this book.

Some material reprinted from "IEEE Std. 1497-2001, IEEE Standard for Standard Delay Format (SDF) for the Electronic Design Process; IEEE Std. 1364-2001, IEEE Standard Verilog Hardware Description Language; IEEE Std.1481-1999, IEEE Standard for Integrated Circuit (IC) Delay and Power Calculation System", with permission from IEEE. The IEEE disclaims any responsibility or liability resulting from the placement and use in the described manner.

Some material reprinted from "LEF/DEF Language Reference, Product Version 5.6" with permission from Cadence Design Systems, Inc.

Some material reprinted from standard cell libraries with permission from ARM Limited, Copyright 2005. This material may be protected by one or more US patents, foreign patents, or pending applications. The material is provided "as is" without warranty of any kind, either express or implied, including but not limited to, the implied warranties of merchantability, satisfactory quality, fitness for a particular purpose or non-infringement.

Copy editing: Linda Zeyak

Printed in the United States of America

10 9 8 7 6 5 4 3 2 1

Library of Congress Control Number: 2005902405

ISBN 0-9650391-3-7

Contents

❏

Preface

Formats, formats and more formats! And it is supposed to be simple, right? One tool writes out a large exchange file of size one Gigabyte and another tool reads this mammoth exchange file and then you find out that the results don't look good. And this is when the arguments start. Which tool is to blame for the error? The file writer or the file reader? And in my experience, quite often, the culprit has been either wrong information in the exchange format file or wrong interpretation by the tool reader. And thus I was forced into the world of exchange formats!

The purpose of this book is to make you, the reader, brave enough to open up one of these huge files without getting intimidated and be able to smartly and quickly understand the contents of one of these exchange format files. The book delves into details of some of the exchange formats that are commonly used in ASIC design. For example, by reading the chapter on Standard Parasitic Exchange Format, you should be able to understand exactly how the capacitance values described in the file and what they mean. And from here it would be easy enough to correlate the value that the tool writes in the file versus what the tool reader interprets and uses for its calculation.

The following exchange formats are described in this book:

i. DEF: Design Exchange Format

ii. LEF: Library Exchange Format

    iii. PDEF: Physical Design Exchange Format

    iv. SDF: Standard Delay Format

    v. SPEF: Standard Parasitic Exchange Format

    vi. VCD: Value Change Dump Format

This book is a primer. It explains the format and high-level details of each of the formats. Intricate details of each of the formats are not described. However, complete syntax for each of the formats is described so that an advanced reader can also use this book as a reference. A reader interested in an in-depth understanding of any of these formats is encouraged to refer to the appropriate official standard document listed in the bibliography.

This book would serve well as a desk reference for all designers: ASIC, FPGA, chip or system designers. Enough information is presented here to get a quick and thorough understanding of each of the formats. All chapters are independent of each other and the reader can flip and jump directly to the chapter of interest.

I would like to acknowledge the excellent and detailed feedback that I have received from Chi-Chang Liaw and Rakesh Chadha at eSilicon Corporation and Mani C. Woodroffe at Synopsys Inc. on earlier versions of this manuscript.

I welcome any feedback that you may have on this book. You can email me at *jbhasker@esilicon.com* or contact me through my publisher.

J. Bhasker
September 2005

## CHAPTER 1

SDF

T his chapter describes the standard delay annotation format and explains how backannotation is performed in simulation. The delay format describes cell delays and interconnect delays of a design netlist and is independent of the language the design may be described in, may it be VHDL or Verilog HDL, the two dominant standard hardware description languages.

1.1   What is it?

SDF stands for Standard Delay Format. It is an IEEE standard - IEEE Std 1497. It is an ASCII text file. It describes timing information and constraints. Its purpose is to serve as a textual timing exchange medium between various tools. It can also be used to describe timing data for tools that require it. Since it is an IEEE standard, timing information generated by one tool can be consumed by a number of other tools that support such a standard. The data is represented in a tool-independent and language-independent way and it includes specification of interconnect delays, device delays and timing checks.

Since SDF is an ASCII file, it is human-readable, though these files tend to be rather large for real designs. However, it is meant as an exchange medium between tools. Quite often when exchanging information, you could potentially run into a problem where a tool generates an SDF file but the other tool that reads SDF does not read the SDF properly. The tool reader could either generate an error or a warning reading the SDF or it might interpret the values in the SDF incorrectly. In that case, you will have to look into the file and see what went wrong. This chapter explains the basics of the SDF file and provides necessary and sufficient information for you to understand and debug any annotation problems.

Figure 1-1 shows a typical flow of how an SDF file is used. A timing calculator tool typically generates the timing information that is stored in an SDF file. This information is then backannotated into the design by the tool that reads the SDF. Note that the complete design information is not captured in an SDF file, but only the delay values are stored. For example, instance names and pin names of instances are captured in the SDF as they are necessary to specify instance-specific or pin-specific delays. Therefore, it is imperative that the same design be presented to both the SDF generation tool and the SDF reader tool.

One design can have multiple SDF files associated with it. One SDF file can be created for one design. In a hierarchical design, multiple SDFs may be created for each block in a hierarchy. During annotation, each SDF is applied to the appropriate hierarchical instance. Figure 1-2 shows this figuratively.

An SDF file contains computed timing data for backannotation and for forward-annotation. More specifically, it contains:

    i.   Cell delays

    ii.   Pulse propagation

    iii. Timing checks

    iv. Interconnect delays

    v.   Timing environment

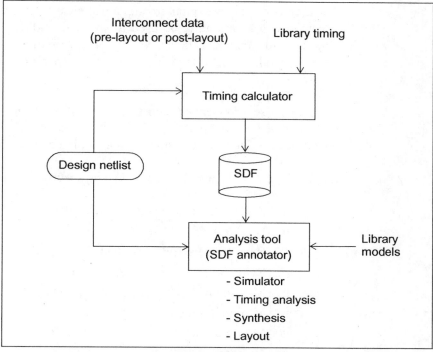

Figure 1-1 The SDF flow.

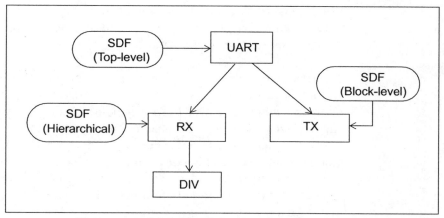

Figure 1-2 Multiple SDFs in a hierarchical design.

Both pin-to-pin delay and distributed delay can be modeled for cell delays. Pin-to-pin delays are represented using the IOPATH construct.

These constructs define input to output path delays for each cell. The COND construct can additionally be used to specify conditional pin-to-pin delays. State-dependent path delays can be specified using the COND construct as well. Distributed delay modeling is specified using the DEVICE construct.

The pulse propagation constructs, PATHPULSE and PATHPULSEPER-CENT, can be used to specify the size of glitches that is allowed to propagate to the output of a cell using the pin-to-pin delay model.

The range of timing checks that can be specified in SDF includes:

i.   Setup: SETUP, SETUPHOLD

ii.   Hold: HOLD, SETUPHOLD

iii. Recovery: RECOVERY, RECREM

iv.   Removal: REMOVAL, RECREM

v.   Maximum skew: SKEW, BIDIRECTSKEW

vi.   Minimum pulse width: WIDTH

vii. Minimum period: PERIOD

viii. No change: NOCHANGE

Conditions may be present on signals in timing checks. Negative values are allowed in timing checks, though tools that don't support negative values can choose to replace it with zero.

There are three styles of interconnect modeling that are supported in an SDF description. The INTERCONNECT construct is the most general and often used and can be used to specify point-to-point delay (from source to sink). Thus a single net can have multiple INTERCONNECT constructs. The PORT construct can be used to specify nets delays at only the load ports - it assumes that there is only one source for the net. The NETDELAY construct can be used to specify the delay of an entire net without regard to the sources or its sinks and therefore is the least specific way of specifying delays on a net.

The timing environment provides information under which the design operates. Such information includes the ARRIVAL, DEPARTURE, SLACK and

WAVEFORM constructs. These constructs are mainly used for forward-annotation, such as for synthesis.

1.2   The Format

An SDF file contains a header section followed by one or more cells. Each cell represents a region or scope in a design. It can be a library primitive or a user-defined black box.

```
(DELAYFILE
  <header_section>
  (CELL
    <cell_section>
  )
  (CELL
    <cell_section>
  )
  ... <other cells>
)
```

The header section contains general information and does not affect the semantics of the SDF file, except for the hierarchy separator, timescale and the SDF version number. The hierarchy separator, DIVIDER, by default is the dot (`.`) character. It can be replaced with the `/` character by specifying:

```
(DIVIDER /)
```

If no timescale information is present in the header, the default is 1ns. Otherwise a timescale, TIMESCALE, can be explicitly specified using:

```
(TIMESCALE 10ps)
```

which says to multiply all delay values specified in the SDF file by 10ps.

The SDF version, SDFVERSION, is required and is used by the consumer of SDF to ensure that the file conforms to the specified SDF version. Other information that may be present in the header section, which is a general information category, includes date, program name, version and operating condition.

```
(DESIGN "BCM")
(DATE "Tuesday, May 24, 2004")
(PROGRAM "Star Galaxy Automation Inc., TimingTool")
(VERSION "V2004.1")
(VOLTAGE 1.65:1.65:1.65)
(PROCESS "1.000:1.000:1.000")
(TEMPERATURE 0.00:0.00:0.00)
```

Following the header section is a description of one or more cells. Each cell represents one or more instances (using wild-card) in the design. A cell may either be a library primitive or a hierarchical block.

```
(CELL
  (CELLTYPE <cell_type>)
  (INSTANCE <hierarchical_instance_name>)
  (DELAY
    <path_delay_section>
  )
  (TIMINGCHECK
    <timing_check_section>
  )
  (TIMINGENV
    <timing_environment_section>
  )
  (LABEL
    <label_section>
  )
)
. . . <other cells>
```

The order of cells is important as data is processed top to bottom. A later cell description may override timing information specified by an earlier cell description (usually it is not common to have timing information of the same cell instance defined twice). In addition, timing information can be annotated either as an absolute value or as an increment. If timing is incrementally applied, it adds the new value to the existing value; if the timing is absolute, it overwrites any previously specified timing information.

The cell instance can be a hierarchical instance name. The separator used for hierarchy separator must conform to the one specified in the header section. The cell instance name can optionally be the `*' character referring to a wildcard character, which means all cell instances of the specified type.

```
(CELL
  (CELLTYPE "NAND2")
  (INSTANCE *)
  // Refers to all instances of NAND2.
  . . .
```

There are four types of timing specifications that can be described in a cell.

i.   DELAY: Used to describe delays.

ii.   TIMINGCHECK: Used to describe timing checks.

iii. TIMINGENV: Used to describe the timing environment.

iv.   LABEL: Declares timing model variables that can be used to describe delays.

Here are some examples.

```
// An absolute path delay specification:
(DELAY
  (ABSOLUTE
    (IOPATH A Y (0.147))
  )
)

// A setup and hold timing check specification:
(TIMINGCHECK
  (SETUPHOLD (posedge Q) (negedge CK) (0.448) (0.412))
)

// A timing constraint between two points:
(TIMINGENV
  (PATHCONSTRAINT UART/ENA UART/TX/CTRL (2.1) (1.5))
)
```

```
// A label that overrides the value of a Verilog HDL
// specparam:
(LABEL
  (ABSOLUTE
    (t$CLK$Q (0.480:0.512:0.578) (0.356:0.399:0.401))
    (tsetup$D$CLK (0.112))
  )
)
```

There are four types of DELAY timing specifications.

i.   ABSOLUTE: Replaces existing delay values for cell instance during backannotation.

ii.   INCREMENT: Adds the new delay data to any existing delay values of the cell instance.

iii.   PATHPULSE: Specifies pulse propagation limit between an input and output of the design. This limit is used to decide whether to propagate a pulse appearing on the input to the output, or to be marked with an 'X', or to get filtered out.

iv.   PATHPULSEPERCENT: This is exactly identical to PATHPULSE except that the values are described as percents.

Here are some examples.

```
// Absolute port delay:
(DELAY
  (ABSOLUTE
    (PORT UART.DIN (0.170))
    (PORT UART.RX.XMIT (0.645))
  )
)

// Adds IO path delay to existing delays of cell:
(DELAY
  (INCREMENT
    (IOPATH (negedge SE) Q (1.1:1.22:1.35))
  )
)
```

```
// Pathpulse delay:
(DELAY
  (PATHPULSE RN Q (3) (7))
)
// The ports RN and Q are input and output of the
// cell. The first value, 3, is the pulse rejection
// limit, called r-limit; it defines the narrowest pulse
// that can appear on output. Any pulse narrower than
// this is rejected, that is, it will not appear on
// output. The second value, 7, if present, is the
// error limit - also called e-limit. Any pulse smaller
// than e-limit causes the output to be an X.
// The e-limit must be greater than r-limit. See
// Figure 1-3.  When a pulse that is less than 3 (r-limit)
```

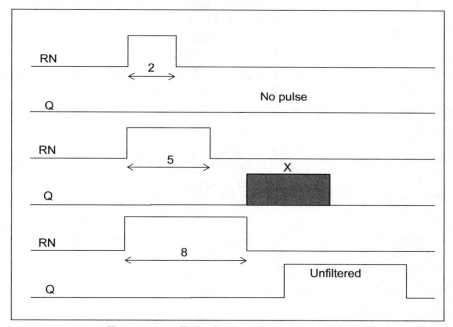

Figure 1-3 Error limit and rejection limit.

```
// occurs, the pulse does not propagate to the output.
// When the pulse width is between the 3 (r-limit) and
// 7 (e-limit), the output is an X. When the pulse width
// is larger than 7 (e-limit), pulse propagates to output
// without any filtering.
```

```
// Pathpulsepercent delay type:
(DELAY
  (PATHPULSEPERCENT CIN SUM (30) (50))
)
// The r-limit is specified as 30% of the delay time from
// CIN to SUM and the e-limit is specified as 50% of
// this delay.
```

There are eight types of delay definitions that can be described with either ABSOLUTE or INCREMENT.

i.   IOPATH: Input-output path delays.

ii.   RETAIN: Retain definition. Specifies the time for which an output shall retain its previous value after a change on its related input port.

iii. COND: Conditional path delay. Can be used to specify state-dependent input-to-output path delays.

iv.   CONDELSE: Default path delay. Specifies default value to use for conditional paths.

v.   PORT: Port delays. Specifies interconnect delays that is modeled as delay at input ports.

vi.   INTERCONNECT: Interconnect delays. Specifies the propagation delay across a net from a source to its sink.

vii. NETDELAY: Net delays. Specifies the propagation delay from all sources to all sinks of a net.

viii. DEVICE: Device delay. Primarily used to describe a distributed timing model. Specifies propagation delay of all paths through a cell to the output port.

Here are some examples.

```
// IO path delay between posedge of CK and Q:
(DELAY
  (ABSOLUTE
    (IOPATH (posedge CK) Q (2) (3))
  )
)
// 2 is the propagation rise delay and 3 is the
// propagation fall delay.
```

```
// Retain delay in an IO path:
(DELAY
  (ABSOLUTE
    (IOPATH A Y
      (RETAIN (0.05:0.05:0.05) (0.04:0.04:0.04))
      (0.101:0.101:0.101) (0.09:0.09:0.09))
  )
)
// Y shall retain its previous value for 50ps (40ps for
// a low value) after a change of value on input A.
// 50ps is the retain high value, 40ps is the retain
// low value, 101ps is the propagate rise delay and
// 90ps is the propagate fall delay. See Figure 1-4.
```

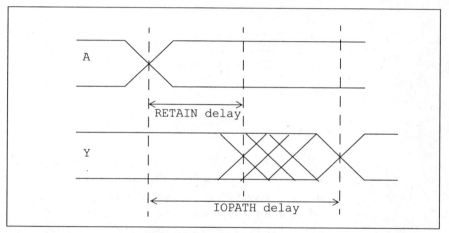

Figure 1-4   RETAIN delay.

```
// Conditional path delay:
(DELAY
  (ABSOLUTE
    (COND SE == 1'b1 (IOPATH (posedge CK) Q (0.661)))
  )
)

// Default conditional path delay:
(DELAY
  (ABSOLUTE
    (CONDELSE (IOPATH ADDR[7] COUNT[0] (0.870) (0.766)))
  )
)
```

```
// Port delay on input FRM_CNT[0]:
(DELAY
  (ABSOLUTE
    (PORT UART/RX/FRM_CNT[0] (0.439))
  )
)

// Interconnect delay:
(DELAY
  (ABSOLUTE
    (INTERCONNECT O1/Y O2/B (0.209:0.209:0.209))
  )
)

// Net delay:
(DELAY
  (ABSOLUTE
    (NETDELAY A3/B (0.566))
  )
)
```

Delays

So far we have seen many different forms of delays. There are additional forms of delay specification. In general, delays can be specified as a set of one, two, three, six or twelve tokens that can be used to describe the following transition delays: 0->1, 1->0, 0->Z, Z->1, 1->Z, Z->0, 0->X, X->1, 1->X, X->0, X->Z, Z->X. The following table shows how fewer than twelve delay tokens are used to represent the twelve transitions.

Transition	*2-values* *(v1 v2)*	*3-values* *(v1 v2 v3)*	*6-values* *(v1 v2 v3* *v4 v5 v6)*	*12-values* *(v1 v2 v3 v4* *v5 v6 v7 v8 v9* *v10 v11 v12)*
0->1	v1	v1	v1	v1
1->0	v2	v2	v2	v2
0->Z	v1	v3	v3	v3

Table 1-5 Mapping to twelve transition delays.

Transition	2-values (v1 v2)	3-values (v1 v2 v3)	6-values (v1 v2 v3 v4 v5 v6)	12-values (v1 v2 v3 v4 v5 v6 v7 v8 v9 v10 v11 v12)
Z->1	v1	v1	v4	v4
1->Z	v2	v3	v5	v5
Z->0	v2	v2	v6	v6
0->X	v1	min(v1,v3)	min(v1,v3)	v7
X->1	v1	v1	max(v1,v4)	v8
1->X	v2	min(v2,v3)	min(v2,v5)	v9
X->0	v2	v2	max(v2,v6)	v10
X->Z	max(v1,v2)	v3	max(v3,v5)	v11
Z->X	min(v1,v2)	min(v1,v2)	min(v6,v4)	v12

Table 1-5 Mapping to twelve transition delays.

Here are some examples of these delays.

```
(DELAY
 (ABSOLUTE
  // 1-value delay:
  (IOPATH A Y (0.989))
  // 2-value delay:
  (IOPATH B Y (0.989) (0.891))
  // 6-value delay:
  (IOPATH CTRL Y (0.121) (0.119) (0.129)
               (0.131) (0.112) (0.124))
  // 12-value delay:
  (COND RN == 1'b0
   (IOPATH C Y (0.330) (0.312) (0.330) (0.311) (0.328)
            (0.321) (0.328) (0.320) (0.320)
            (0.318) (0.318) (0.316)
   )
  )
  // In this 2-value delay, the first one is null
  // implying the annotator is not to change its value.
  (IOPATH RN Q () (0.129))
```

```
      )
    )
```

Each delay token can, in turn, be written as one, two or three values as shown in the following examples.

```
(DELAY
  (ABSOLUTE
    // One value in a delay token:
    (IOPATH A Y (0.117))
    // The delay value, the pulse rejection limit
    // (r-limit) and X filter limit (e-limit) are same.

    // Two values in a delay (note no colon):
    (IOPATH (posedge CK) Q (0.12 0.15))
    // 0.12 is the delay value and 0.15 is the r-limit
    // and e-limit.

    // Three values in a delay:
    (IOPATH F1/Y AND1/A (0.339 0.1 0.15))
    // Path delay is 0.339, r-limit is 0.1 and
    // e-limit is 0.15.
  )
)
```

Delay values in a single SDF file can be written using signed real numbers or as triplets of form:

```
(8.0 : 3.6 : 9.8)
```

to denote minimum, typical, maximum delays that represent the three process operating conditions of the design. The choice of which value is selected is made by the annotator typically based on a user-provided option. The values in the triplet form are optional, though it should have at least one. For example, the following are legal.

```
(::0.22)
(1.001: :0.998)
```

Values that are not specified are simply not annotated.

Timing Checks

Timing check limits are specified in the section that starts with the TIMINGCHECK keyword. In any of these checks, a COND construct can be used to specify conditional timing checks. In some cases, two additional conditional checks can be specified, SCOND and CCOND, that are associated with the *stamp event* and the *check event*.

Following are the set of checks:

- *i.* SETUP: Setup timing check
- *ii.* HOLD: Hold timing check
- *iii.* SETUPHOLD: Setup and hold timing check
- *iv.* RECOVERY: Recovery timing check
- *v.* REMOVAL: Removal timing check
- *vi.* RECREM: Recovery and removal timing check
- *vii.* SKEW: Unidirectional skew timing check
- *viii.* BIDIRECTSKEW: Bidirectional skew timing check
- *ix.* WIDTH: Width timing check
- *x.* PERIOD: Period timing check
- *xi.* NOCHANGE: Nochange timing check

Here are some examples.

```
(TIMINGCHECK
  // Setup check limit:
  (SETUP din (posedge clk) (2))

  // Hold check limit:
  (HOLD din (negedge clk) (0.445:0.445:0.445))

  // Conditional hold check limit :
  (HOLD (COND RST==1'b1 D) (posedge CLK) (1.15))
  // Hold check between D and positive edge of CLK, but
  // only when RST is 1.

  // Setup and hold check limit:
  (SETUPHOLD J CLK (1.2) (0.99))
  // 1.2 is the setup limit and 0.99 is the hold limit.
```

```
// Conditional setup and hold limit:
(SETUPHOLD D CLK (0.809) (0.591) (CCOND ~SE))
// Condition applies with CLK for setup and
// with D for hold.

// Conditional setup and hold check limit:
(SETUPHOLD (COND ~RST D) (posedge CLK) (1.452) (1.11))
// Setup and hold check between D and positive edge
// of CLK, but only when RST is low.

// RECOVERY check limit:
(RECOVERY SE (negedge CLK) (0.671))

// Conditional removal check limit:
(REMOVAL (COND ~LOAD CLEAR) CLK (2.001:2.1:2.145))
// Removal check between CLEAR and CLK but only
// when LOAD is low.

// Recovery and removal check limit:
(RECREM RST (negedge CLK) (1.1) (0.701))
// 1.1 is the recovery limit and 0.701 is the
// removal limit.

// Skew conditional check limit:
(SKEW (COND MMODE==1'b1 GNT) (posedge REQ) (3.2))

// Bidirectional skew check limit:
(BIDIRECTSKEW (posedge CLOCK1) (negedge TCK) (1.409))

// Width check limit:
(WIDTH (negedge RST) (12))

// Period check limit:
(PERIOD (posedge JTCLK) (13.33))

// Nochange check limit:
(NOCHANGE (posedge REQ) (negedge GNT) (2.5) (3.12))
)
```

Labels

Labels are used to specify values for VHDL generics or Verilog HDL specify parameters.

```
(LABEL
  (ABSOLUTE
    (thold$d$clk (0.809))
    (tph$A$Y (0.553))
  )
)
```

Timing Environment

There are a number of constructs available that can be used to describe the timing environment of a design. However, these constructs are used for forward-annotation rather than backward-annotation, such as in logic synthesis tools. These are not described in this text.

1.2.1   Examples

We provide complete SDFs for two designs.

Full-adder

Here is the Verilog HDL netlist for a full-adder circuit.

```
module FA_STR (A, B, CIN, SUM, COUT);
  input A, B, CIN;
  output SUM, COUT;
  wire S1, S2, S3, S4, S5;

  XOR2X1 X1 (.Y(S1), .A(A), .B(B));
  XOR2X1 X2 (.Y(SUM), .A(S1), .B(CIN));

  AND2X1 A1 (.Y(S2), .A(A), .B(B));
  AND2X1 A2 (.Y(S3), .A(B), .B(CIN));
  AND2X1 A3 (.Y(S4), .A(A), .B(CIN));

  OR2X1 O1 (.Y(S5), .A(S2), .B(S3));
```

```
    OR2X1 O2 (.Y(COUT), .A(S4), .B(S5));
endmodule
```

Here is the complete corresponding SDF file produced by a timing analysis tool.

```
(DELAYFILE
  (SDFVERSION "OVI 2.1")
  (DESIGN "FA_STR")
  (DATE "Mon May 24 13:56:43 2004")
  (VENDOR "slow")
  (PROGRAM "CompanyName ToolName")
  (VERSION "V2.3")
  (DIVIDER /)
  // OPERATING CONDITION "slow"
  (VOLTAGE 1.35:1.35:1.35)
  (PROCESS "1.000:1.000:1.000")
  (TEMPERATURE 125.00:125.00:125.00)
  (TIMESCALE 1ns)
  (CELL
    (CELLTYPE "FA_STR")
    (INSTANCE)
    (DELAY
      (ABSOLUTE
        (INTERCONNECT A A3/A (0.000:0.000:0.000))
        (INTERCONNECT A A1/A (0.000:0.000:0.000))
        (INTERCONNECT A X1/A (0.000:0.000:0.000))
        (INTERCONNECT B A2/A (0.000:0.000:0.000))
        (INTERCONNECT B A1/B (0.000:0.000:0.000))
        (INTERCONNECT B X1/B (0.000:0.000:0.000))
        (INTERCONNECT CIN A3/B (0.000:0.000:0.000))
        (INTERCONNECT CIN A2/B (0.000:0.000:0.000))
        (INTERCONNECT CIN X2/B (0.000:0.000:0.000))
        (INTERCONNECT X2/Y SUM (0.000:0.000:0.000))
        (INTERCONNECT O2/Y COUT (0.000:0.000:0.000))
        (INTERCONNECT X1/Y X2/A (0.000:0.000:0.000))
        (INTERCONNECT A1/Y O1/A (0.000:0.000:0.000))
        (INTERCONNECT A2/Y O1/B (0.000:0.000:0.000))
        (INTERCONNECT A3/Y O2/A (0.000:0.000:0.000))
        (INTERCONNECT O1/Y O2/B (0.000:0.000:0.000))
      )
    )
  )
)
```

```
(CELL
  (CELLTYPE "XOR2X1")
  (INSTANCE X1)
  (DELAY
    (ABSOLUTE
      (IOPATH A Y (0.197:0.197:0.197)
                  (0.190:0.190:0.190))
      (IOPATH B Y (0.209:0.209:0.209)
                  (0.227:0.227:0.227))
      (COND B==1'b1 (IOPATH A Y (0.197:0.197:0.197)
                                (0.190:0.190:0.190)))
      (COND A==1'b1 (IOPATH B Y (0.209:0.209:0.209)
                                (0.227:0.227:0.227)))
      (COND B==1'b0 (IOPATH A Y (0.134:0.134:0.134)
                                (0.137:0.137:0.137)))
      (COND A==1'b0 (IOPATH B Y (0.150:0.150:0.150)
                                (0.163:0.163:0.163)))
    )
  )
)
(CELL
  (CELLTYPE "XOR2X1")
  (INSTANCE X2)
  (DELAY
    (ABSOLUTE
      (IOPATH (posedge A) Y (0.204:0.204:0.204)
                            (0.196:0.196:0.196))
      (IOPATH (negedge A) Y (0.198:0.198:0.198)
                            (0.190:0.190:0.190))
      (IOPATH B Y (0.181:0.181:0.181)
                  (0.201:0.201:0.201))
      (COND B==1'b1 (IOPATH A Y (0.198:0.198:0.198)
                                (0.196:0.196:0.196)))
      (COND A==1'b1 (IOPATH B Y (0.181:0.181:0.181)
                                (0.201:0.201:0.201)))
      (COND B==1'b0 (IOPATH A Y (0.135:0.135:0.135)
                                (0.140:0.140:0.140)))
      (COND A==1'b0 (IOPATH B Y (0.122:0.122:0.122)
                                (0.139:0.139:0.139)))
    )
  )
)
```

```
(CELL
  (CELLTYPE "AND2X1")
  (INSTANCE A1)
  (DELAY
    (ABSOLUTE
      (IOPATH A Y (0.147:0.147:0.147)
                  (0.157:0.157:0.157))
      (IOPATH B Y (0.159:0.159:0.159)
                  (0.173:0.173:0.173))
    )
  )
)
(CELL
  (CELLTYPE "AND2X1")
  (INSTANCE A2)
  (DELAY
    (ABSOLUTE
      (IOPATH A Y (0.148:0.148:0.148)
                  (0.157:0.157:0.157))
      (IOPATH B Y (0.160:0.160:0.160)
                  (0.174:0.174:0.174))
    )
  )
)
(CELL
  (CELLTYPE "AND2X1")
  (INSTANCE A3)
  (DELAY
    (ABSOLUTE
      (IOPATH A Y (0.147:0.147:0.147)
                  (0.157:0.157:0.157))
      (IOPATH B Y (0.159:0.159:0.159)
                  (0.173:0.173:0.173))
    )
  )
)
(CELL
  (CELLTYPE "OR2X1")
  (INSTANCE O1)
  (DELAY
    (ABSOLUTE
      (IOPATH A Y (0.138:0.138:0.138)
                  (0.203:0.203:0.203))
```

```
(IOPATH B Y (0.151:0.151:0.151)
             (0.223:0.223:0.223))
  )
 )
)
(CELL
  (CELLTYPE "OR2X1")
  (INSTANCE O2)
  (DELAY
    (ABSOLUTE
      (IOPATH A Y (0.126:0.126:0.126)
                   (0.191:0.191:0.191))
      (IOPATH B Y (0.136:0.136:0.136)
                   (0.212:0.212:0.212))
    )
   )
  )
)
```

All delays in the INTERCONNECTs are 0 as this is prelayout data and ideal interconnects are modeled.

Here are the Verilog HDL specify blocks[1] for the cell types used in the full-adder.

```
// Specify block for OR2X1:
specify
  // Delay parameters:
  specparam
    tplh$A$Y = 1.0,
    tphl$A$Y = 1.0,
    tplh$B$Y = 1.0,
    tphl$B$Y = 1.0;

    // Path delays:
    (A *> Y) = (tplh$A$Y, tphl$A$Y);
    (B *> Y) = (tplh$B$Y, tphl$B$Y);
endspecify
```

---

1. The specify blocks are reprinted from standard cell libraries with permission from ARM Limited, Copyright 2005. All rights reserved.

```
// Specify block for XOR2X1:
specify
  // Delay parameters:
  specparam
    tplh$A$Y = 1.0,
    tphl$A$Y = 1.0,
    tplh$B$Y = 1.0,
    tphl$B$Y = 1.0;

  // Path delays:
  if (B == 1'b1)
    (A *> Y) = (tplh$A$Y, tphl$A$Y);
  if (B == 1'b0)
    (A *> Y) = (tplh$A$Y, tphl$A$Y);
  if (A == 1'b1)
    (B *> Y) = (tplh$B$Y, tphl$B$Y);
  if (A == 1'b0)
    (B *> Y) = (tplh$B$Y, tphl$B$Y);
endspecify

// Specify block for AND2X1:
specify
  // Delay parameters:
  specparam
    tplh$A$Y = 1.0,
    tphl$A$Y = 1.0,
    tplh$B$Y = 1.0,
    tphl$B$Y = 1.0;

  // Path delays:
  (A *> Y) = (tplh$A$Y, tphl$A$Y);
  (B *> Y) = (tplh$B$Y, tphl$B$Y);
endspecify
```

Here are the VHDL generics[1] declared for the cell types used in the full-adder.

```
-- Generic declarations for AND2X1:
generic (
```

---

1. The generic declarations are reprinted from standard cell libraries with permission from ARM Limited, Copyright 2005. All rights reserved.

```
    XOn             : BOOLEAN := DefCombSpikeXOn;
    MsgOn           : BOOLEAN := DefCombSpikeMsgOn;
    InstancePath : STRING  := "*";
    tipd_A          : VitalDelayType01
                        := (DefDummyIpd, DefDummyIpd);
    tpd_A_Y         : VitalDelayType01
                        := (DefDummyDelay, DefDummyDelay);
    tipd_B          : VitalDelayType01
                        := (DefDummyIpd, DefDummyIpd);
    tpd_B_Y         : VitalDelayType01
                        := (DefDummyDelay, DefDummyDelay)
);

-- Generic declaration for XOR2X1:
generic (
  XOn             : BOOLEAN := DefCombSpikeXOn;
  MsgOn           : BOOLEAN := DefCombSpikeMsgOn;
  InstancePath : STRING := "*";
  tipd_A          : VitalDelayType01
                      := (DefDummyIpd, DefDummyIpd);
  tipd_B          : VitalDelayType01
                      := (DefDummyIpd, DefDummyIpd);
  tpd_A_Y_B_EQ_0 : VitalDelayType01
                      := (DefDummyDelay, DefDummyDelay);
  tpd_A_Y_B_EQ_1 : VitalDelayType01
                      := (DefDummyDelay, DefDummyDelay);
  tpd_A_Y         : VitalDelayType01
                      := (DefDummyDelay, DefDummyDelay);
  tpd_B_Y_A_EQ_0 : VitalDelayType01
                      := (DefDummyDelay, DefDummyDelay);
  tpd_B_Y_A_EQ_1 : VitalDelayType01
                      := (DefDummyDelay, DefDummyDelay);
  tpd_B_Y         : VitalDelayType01
                      := (DefDummyDelay, DefDummyDelay)
);

-- Generic declarations for OR2X1:
generic (
  XOn             : BOOLEAN := DefCombSpikeXOn;
  MsgOn           : BOOLEAN := DefCombSpikeMsgOn;
  InstancePath : STRING  := "*";
  tipd_A          : VitalDelayType01
                      := (DefDummyIpd, DefDummyIpd);
```

```
    tpd_A_Y          : VitalDelayType01
                        := (DefDummyDelay, DefDummyDelay);
    tipd_B           : VitalDelayType01
                        := (DefDummyIpd, DefDummyIpd);
    tpd_B_Y          : VitalDelayType01
                        := (DefDummyDelay, DefDummyDelay)
);
```

Decade Counter

Here is the Verilog HDL model for a decade counter.

```
module DECADE_CTR (COUNT, Z);
  input COUNT;
  output [0:3] Z;
  wire S1, S2;

  AND2X1 a1 (.Y(S1), .A(Z[2]), .B(Z[1]));

  JKFFX1
    JK1 (.J(1'b1), .K(1'b1), .CK(COUNT),
          .Q(Z[0]), .QN()),
    JK2 (.J(S2), .K(1'b1), .CK(Z[0]), .Q(Z[1]), .QN()),
    JK3 (.J(1'b1), .K(1'b1), .CK(Z[1]),
          .Q(Z[2]), .QN()),
    JK4 (.J(S1), .K(1'b1), .CK(Z[0]),
          .Q(Z[3]), .QN(S2));
endmodule
```

The complete corresponding SDF follows.

```
(DELAYFILE
  (SDFVERSION "OVI 2.1")
  (DESIGN "DECADE_CTR")
  (DATE "Mon May 24 14:30:17 2004")
  (VENDOR "Star Galaxy Automation, Inc.")
  (PROGRAM "MyCompanyName ToolTime")
  (VERSION "V2.3")
  (DIVIDER /)
  // OPERATING CONDITION "slow"
  (VOLTAGE 1.35:1.35:1.35)
  (PROCESS "1.000:1.000:1.000")
```

```
(TEMPERATURE 125.00:125.00:125.00)
(TIMESCALE 1ns)
(CELL
  (CELLTYPE "DECADE_CTR")
  (INSTANCE)
  (DELAY
    (ABSOLUTE
      (INTERCONNECT COUNT JK1/CK (0.191:0.191:0.191))
      (INTERCONNECT JK1/Q Z\[0\] (0.252:0.252:0.252))
      (INTERCONNECT JK2/Q Z\[1\] (0.186:0.186:0.186))
      (INTERCONNECT JK3/Q Z\[2\] (0.18:0.18:0.18))
      (INTERCONNECT JK4/Q Z\[3\] (0.195:0.195:0.195))
      (INTERCONNECT JK3/Q a1/A (0.175:0.175:0.175))
      (INTERCONNECT JK2/Q a1/B (0.207:0.207:0.207))
      (INTERCONNECT JK4/QN JK2/J (0.22:0.22:0.22))
      (INTERCONNECT JK1/Q JK2/CK (0.181:0.181:0.181))
      (INTERCONNECT JK2/Q JK3/CK (0.193:0.193:0.193))
      (INTERCONNECT a1/Y JK4/J (0.224:0.224:0.224))
      (INTERCONNECT JK1/Q JK4/CK (0.218:0.218:0.218))
    )
  )
)
(CELL
  (CELLTYPE "AND2X1")
  (INSTANCE a1)
  (DELAY
    (ABSOLUTE
      (IOPATH A Y (0.179:0.179:0.179)
                  (0.186:0.186:0.186))
      (IOPATH B Y (0.190:0.190:0.190)
                  (0.210:0.210:0.210))
    )
  )
)
(CELL
  (CELLTYPE "JKFFX1")
  (INSTANCE JK1)
  (DELAY
    (ABSOLUTE
      (IOPATH (posedge CK) Q (0.369:0.369:0.369)
                             (0.470:0.470:0.470))
      (IOPATH (posedge CK) QN (0.280:0.280:0.280)
                              (0.178:0.178:0.178))
```

```
        )
      )
      (TIMINGCHECK
        (SETUP (posedge J) (posedge CK)
          (0.362:0.362:0.362))
        (SETUP (negedge J) (posedge CK)
          (0.220:0.220:0.220))
        (HOLD (posedge J) (posedge CK)
          (-0.272:-0.272:-0.272))
        (HOLD (negedge J) (posedge CK)
          (-0.200:-0.200:-0.200))
        (SETUP (posedge K) (posedge CK)
          (0.170:0.170:0.170))
        (SETUP (negedge K) (posedge CK)
          (0.478:0.478:0.478))
        (HOLD (posedge K) (posedge CK)
          (-0.158:-0.158:-0.158))
        (HOLD (negedge K) (posedge CK)
          (-0.417:-0.417:-0.417))
        (WIDTH (negedge CK)
          (0.337:0.337:0.337))
        (WIDTH (posedge CK) (0.148:0.148:0.148))
      )
    )
    (CELL
      (CELLTYPE "JKFFX1")
      (INSTANCE JK2)
      (DELAY
        (ABSOLUTE
          (IOPATH (posedge CK) Q (0.409:0.409:0.409)
                                 (0.512:0.512:0.512))
          (IOPATH (posedge CK) QN (0.326:0.326:0.326)
                                  (0.222:0.222:0.222))
        )
      )
      (TIMINGCHECK
        (SETUP (posedge J) (posedge CK)
          (0.348:0.348:0.348))
        (SETUP (negedge J) (posedge CK)
          (0.227:0.227:0.227))
        (HOLD (posedge J) (posedge CK)
          (-0.257:-0.257:-0.257))
        (HOLD (negedge J) (posedge CK)
```

```
             (-0.209:-0.209:-0.209))
        (SETUP (posedge K) (posedge CK)
         (0.163:0.163:0.163))
        (SETUP (negedge K) (posedge CK)
         (0.448:0.448:0.448))
        (HOLD (posedge K) (posedge CK)
         (-0.151:-0.151:-0.151))
        (HOLD (negedge K) (posedge CK)
         (-0.392:-0.392:-0.392))
        (WIDTH (negedge CK) (0.337:0.337:0.337))
        (WIDTH (posedge CK) (0.148:0.148:0.148))
      )
   )
 (CELL
   (CELLTYPE "JKFFX1")
   (INSTANCE JK3)
   (DELAY
     (ABSOLUTE
       (IOPATH (posedge CK) Q (0.378:0.378:0.378)
                             (0.485:0.485:0.485))
       (IOPATH (posedge CK) QN (0.324:0.324:0.324)
                               (0.221:0.221:0.221))
     )
   )
   (TIMINGCHECK
     (SETUP (posedge J) (posedge CK)
      (0.339:0.339:0.339))
     (SETUP (negedge J) (posedge CK)
      (0.211:0.211:0.211))
     (HOLD (posedge J) (posedge CK)
      (-0.249:-0.249:-0.249))
     (HOLD (negedge J) (posedge CK)
      (-0.192:-0.192:-0.192))
     (SETUP (posedge K) (posedge CK)
      (0.163:0.163:0.163))
     (SETUP (negedge K) (posedge CK)
      (0.449:0.449:0.449))
     (HOLD (posedge K) (posedge CK)
      (-0.152:-0.152:-0.152))
     (HOLD (negedge K) (posedge CK)
      (-0.393:-0.393:-0.393))
     (WIDTH (negedge CK) (0.337:0.337:0.337))
     (WIDTH (posedge CK) (0.148:0.148:0.148))
```

```
        )
      )
      (CELL
        (CELLTYPE "JKFFX1")
        (INSTANCE JK4)
        (DELAY
          (ABSOLUTE
            (IOPATH (posedge CK) Q (0.354:0.354:0.354)
                                   (0.464:0.464:0.464))
            (IOPATH (posedge CK) QN (0.364:0.364:0.364)
                                    (0.256:0.256:0.256))
          )
        )
        (TIMINGCHECK
          (SETUP (posedge J) (posedge CK)
            (0.347:0.347:0.347))
          (SETUP (negedge J) (posedge CK)
            (0.226:0.226:0.226))
          (HOLD (posedge J) (posedge CK)
            (-0.256:-0.256:-0.256))
          (HOLD (negedge J) (posedge CK)
            (-0.208:-0.208:-0.208))
          (SETUP (posedge K) (posedge CK)
            (0.163:0.163:0.163))
          (SETUP (negedge K) (posedge CK)
            (0.448:0.448:0.448))
          (HOLD (posedge K) (posedge CK)
            (-0.151:-0.151:-0.151))
          (HOLD (negedge K) (posedge CK)
            (-0.392:-0.392:-0.392))
          (WIDTH (negedge CK) (0.337:0.337:0.337))
          (WIDTH (posedge CK) (0.148:0.148:0.148))
        )
      )
    )
```

Here are the Verilog HDL specify blocks[1] for the cell types used in the decade counter.

---

1. The specify blocks are reprinted from standard cell libraries with permission from ARM Limited, Copyright 2005. All rights reserved.

```
// Specify block for AND2X1 is described in the
// previous section.

// Specify block for JKFFX1:
specify
  specparam
    // Timing parameters:
    tplh$CK$Q      = 1.0,
    tphl$CK$Q      = 1.0,
    tplh$CK$QN     = 1.0,
    tphl$CK$QN     = 1.0,
    tsetup$J$CK    = 1.0,
    thold$J$CK     = 1.0,
    tsetup$K$CK    = 1.0,
    thold$K$CK     = 1.0,
    tminpwl$CK     = 1.0,
    tminpwh$CK     = 1.0;

    // Path delays:
    if (SandRandJandKb)
      (posedge CK *> (Q    +: 1'b1)) = (tplh$CK$Q);
    if (SandRandJbandK)
      (posedge CK *> (Q    +: 1'b0)) = (tphl$CK$Q);
    if (SandRandJandK)
      (CK *> Q) = (tplh$CK$Q , tphl$CK$Q);
    if (SandRandJandKb)
      (posedge CK *> (QN    +: 1'b0)) = (tphl$CK$QN);
    if (SandRandJbandK)
      (posedge CK *> (QN    +: 1'b1)) = (tplh$CK$QN);
    if (SandRandJandK)
      (CK *> QN) = (tplh$CK$QN , tphl$CK$QN);
    $setuphold (posedge CK && (SandR == 1), posedge J,
      tsetup$J$CK, thold$J$CK, NOTIFIER);
    $setuphold (posedge CK && (SandR == 1), negedge J,
      tsetup$J$CK, thold$J$CK, NOTIFIER);
    $setuphold (posedge CK && (SandR == 1), posedge K,
      tsetup$K$CK, thold$K$CK, NOTIFIER);
    $setuphold (posedge CK && (SandR == 1), negedge K,
      tsetup$K$CK, thold$K$CK, NOTIFIER);
    $width (negedge CK && (SandR == 1), tminpwl$CK, 0,
      NOTIFIER);
    $width (posedge CK && (SandR == 1), tminpwh$CK, 0,
```

```
        NOTIFIER);
endspecify
```

Here are the VHDL generic declarations[1] for the cell types used in the decade counter.

```
-- Generic declaration for AND2X1 is described in the
-- previous section.

-- Generic declarations for JKFFX1:
generic (
  tipd_J      : VitalDelayType01
    := (DefDummyIpd, DefDummyIpd);
  tipd_K      : VitalDelayType01
    := (DefDummyIpd, DefDummyIpd);
  tipd_CK     : VitalDelayType01
    := (DefDummyIpd, DefDummyIpd);
  tisd_J_CK   : VitalDelayType := DefDummyIsd;
  tisd_K_CK   : VitalDelayType := DefDummyIsd;
  ticd_CK     : VitalDelayType := DefDummyIcd;
  tpd_CK_Q    : VitalDelayType01
    := (DefDummyDelay, DefDummyDelay);
  tpd_CK_QN   : VitalDelayType01
    := (DefDummyDelay, DefDummyDelay);
  tsetup_J_CK_posedge_posedge   : VitalDelayType
    := DefDummySetup;
  tsetup_J_CK_negedge_posedge   : VitalDelayType
    := DefDummySetup;
  thold_J_CK_posedge_posedge    : VitalDelayType
    := DefDummyHold;
  thold_J_CK_negedge_posedge    : VitalDelayType
    := DefDummyHold;
  tsetup_K_CK_posedge_posedge   : VitalDelayType
    := DefDummySetup;
  tsetup_K_CK_negedge_posedge   : VitalDelayType
    := DefDummySetup;
  thold_K_CK_posedge_posedge    : VitalDelayType
    := DefDummyHold;
  thold_K_CK_negedge_posedge    : VitalDelayType
```

---

1. The generic declarations are reprinted from standard cell libraries with permission from ARM Limited, Copyright 2005. All rights reserved.

```
                    := DefDummyHold;
        tpw_CK_negedge : VitalDelayType := DefDummyWidth;
        tpw_CK_posedge : VitalDelayType := DefDummyWidth;
        TimingChecksOn : BOOLEAN := false;
        XOn            : Boolean := DefCombSpikeXOn;
        MsgOn          : Boolean := DefCombSpikeMsgOn;
        instancePath   : STRING := "*"
    );
```

1.3   The Annotation Process

In this section we describe how the annotation of the SDF occurs to an HDL description. SDF annotation can be performed by a number of tools, such as logic synthesis, simulation and static timing analysis; the SDF annotator is the component of these tools that reads the SDF, interprets and annotates the timing values to the design. It is assumed that the SDF file is created using information that is consistent with the HDL model and that the same HDL model is used during backannotation. Additionally, it is the responsibility of the SDF annotator to ensure that the timing values in the SDF are interpreted correctly.

The SDF annotator annotates the backannotation timing generics and parameters. It reports any errors if there is any noncompliance to the standard, either in syntax or in the mapping process. If certain SDF constructs are not supported by an SDF annotator, no errors are produced - the annotator simply ignores these.

If the SDF annotator fails to modify a backannotation timing generic, then the value of the generic is not modified during the backannotation process, that is, it is left unchanged.

In a simulation tool, backannotation typically occurs just following the elaboration phase and directly preceding negative constraint delay calculation.

1.3.1   Verilog HDL

In Verilog HDL, the primary mechanism for annotation is the specify block. A specify block can specify path delays and timing checks. Actual delay values and timing check limit values are specified via the SDF file. The mapping is an industry-standard and is defined in IEEE Std 1364.

Specify path delays, specparam values, timing check constraint limits and interconnect delays are among the information obtained from an SDF file and annotated in a specify block of a Verilog HDL module. Other constructs in an SDF file are ignored when annotating to a Verilog HDL model. The LABEL section in SDF defines specparam values. Backannotation is done by matching SDF constructs to corresponding Verilog HDL declarations and then replacing the existing timing values with those in the SDF file.

Here is a table that shows how SDF delay values are mapped to Verilog HDL delay values.

Verilog transition	1-value (v1)	2-values (v1 v2)	3-values (v1 v2 v3)	6-values (v1 v2 v3 v4 v5 v6)	12-values (v1 v2 v3 v4 v5 v6 v7 v8 v9 v10 v11 v12)
0->1	v1	v1	v1	v1	v1
1->0	v1	v2	v2	v2	v2
0->z	v1	v1	v3	v3	v3
z->1	v1	v1	v1	v4	v4
1->z	v1	v2	v3	v5	v5
z->0	v1	v2	v2	v6	v6
0->x	v1	v1	min(v1 v3)	min(v1 v3)	v7
x->1	v1	v1	v1	max(v1 v4)	v8
1->x	v1	v2	min (v2 v3)	min(v2 v5)	v9

Table 1-6 Mapping SDF delays to Verilog HDL delays.

Verilog transition	1-value (v1)	2-values (v1 v2)	3-values (v1 v2 v3)	6-values (v1 v2 v3 v4 v5 v6)	12-values (v1 v2 v3 v4 v5 v6 v7 v8 v9 v10 v11 v12)
x->0	v1	v2	v2	max(v2 v6)	v10
x->z	v1	max(v1 v2)	v3	max(v3 v5)	v11
z->x	v1	min(v1 v2)	min(v1 v2)	min(v4 v6)	v12

Table 1-6 Mapping SDF delays to Verilog HDL delays.

The following table describes the mapping of SDF constructs to Verilog HDL constructs.

Kinds	SDF construct	Verilog HDL
Propagation delay	IOPATH	Specify paths
Input setup time	SETUP	**$setup, $setuphold**
Input hold time	HOLD	**$hold, $setuphold**
Input setup and hold	SETUPHOLD	**$setup, $hold, $setuphold**
Input recovery time	RECOVERY	**$recovery**
Input removal time	REMOVAL	**$removal**
Recovery and removal	RECREM	**$recovery, $removal, $recrem**
Period	PERIOD	**$period**
Pulse width	WIDTH	**$width**
Input skew time	SKEW	**$skew**
No-change time	NOCHANGE	**$nochange**
Port delay	PORT	Interconnect delay
Net delay	NETDELAY	Interconnect delay

Table 1-7 Mapping of SDF to Verilog HDL.

Kinds	SDF construct	Verilog HDL
Interconnect delay	INTERCONNECT	Interconnect delay
Device delay	DEVICE_DELAY	Specify paths
Path pulse limit	PATHPULSE	Specify path pulse limit
Path pulse limit	PATHPULSEPERCENT	Specify path pulse limit

Table 1-7 Mapping of SDF to Verilog HDL.

See later section for examples.

1.3.2   VHDL

Annotation of SDF to VHDL is an industry-standard. It is defined in the IEEE standard for VITAL ASIC Modeling Specification, IEEE Std 1076.4; one of the components of this standard describes the annotation of SDF delays into ASIC libraries. Here, we present only the relevant part of the VITAL standard as it relates to SDF mapping.

· SDF is used to modify backannotation timing generics in a VITAL-compliant model directly. Timing data can be specified only for a VITAL-compliant model using SDF. There are two ways to pass timing data into a VHDL model: via configurations, or directly into simulation. The SDF annotation process consists of mapping SDF constructs and corresponding generics in a VITAL-compliant model during simulation.

In a VITAL-compliant model, there are rules on how generics are to be named and declared that ensures that a mapping can be established between the timing generics of a model and the corresponding SDF timing information.

A timing generic is made up of a generic name and its type. The name specifies the kind of timing information and the type of the generic specifies the kind of timing value. If the name of generic does not follow the VITAL standard, then it is not a timing generic and does not get annotated.

Here is the table showing how SDF delay values are mapped to VHDL delays.

VHDL transition	1-value (v1)	2-values (v1 v2)	3-values (v1 v2 v3)	6-values (v1 v2 v3 v4 v5 v6)	12-values (v1 v2 v3 v4 v5 v6 v7 v8 v9 v10 v11 v12)
0->1	v1	v1	v1	v1	v1
1->0	v1	v2	v2	v2	v2
0->z	v1	v1	v3	v3	v3
z->1	v1	v1	v1	v4	v4
1->z	v1	v2	v3	v5	v5
z->0	v1	v2	v2	v6	v6
0->x	-	-	-	-	v7
x->1	-	-	-	-	v8
1->x	-	-	-	-	v9
x->0	-	-	-	-	v10
x->z	-	-	-	-	v11
z->x	-	-	-	-	v12

Table 1-8 Mapping SDF delays to VHDL delays.

In VHDL, timing information is backannotated via generics. Generic names follow a certain convention so as to be consistent or derived from SDF constructs. With each of the timing generic names, an optional suffix of a conditioned edge can be specified. The edge specifies an edge associated with the timing information. Table 1-9 shows the different kinds of timing generic names.

Kinds	SDF construct	VHDL generic
Propagation delay	IOPATH	**tpd**_InputPort_OutputPort [_condition]
Input setup time	SETUP	**tsetup**_TestPort_RefPort [_condition]

Table 1-9 Mapping of SDF to VHDL generics.

Kinds	SDF construct	VHDL generic
Input hold time	HOLD	**thold**_*TestPort_RefPort* [_*condition*]
Input recovery time	RECOVERY	**trecovery**_*TestPort_RefPort* [_*condition*]
Input removal time	REMOVAL	**tremoval**_*TestPort_RefPort* [_*condition*]
Period	PERIOD	**tperiod**_*InputPort* [_*condition*]
Pulse width	WIDTH	**tpw**_*InputPort* [_*condition*]
Input skew time	SKEW	**tskew**_*FirstPort_SecondPort* [_*condition*]
No-change time	NOCHANGE	**tncsetup**_*TestPort_RefPort* [_*condition*] **tnchold**_*TestPort_RefPort* [_*condition*]
Interconnect path delay	PORT	**tipd**_*InputPort*
Device delay	DEVICE	**tdevice**_*InstanceName* [_*OutputPort*]
Internal signal delay		**tisd**_*InputPort_ClockPort*
Biased propagation delay		**tbpd**_*InputPort_OutputPort_ClockPort* [_*condition*]
Internal clock delay		**ticd**_*ClockPort*

Table 1-9 Mapping of SDF to VHDL generics.

1.4   Mapping Examples

Here are examples of mapping SDF constructs to VHDL generics and Verilog HDL declarations.

Propagation Delay

- Propagation delay from input port A to output port Y with a rise time of 0.406 and a fall of 0.339.

```
// SDF:
(IOPATH A Y (0.406) (0.339))

-- VHDL generic:
tpd_A_Y : VitalDelayType01;
```

```
// Verilog HDL specify path:
(A *> Y) = (tplh$A$Y, tphl$A$Y);
```

- Propagation delay from input port OE to output port Y with a rise time of 0.441 and a fall of 0.409. The minimum, nominal and maximum delays are identical.

```
// SDF:
(IOPATH OE Y (0.441:0.441:0.441) (0.409:0.409:0.409))

-- VHDL generic:
tpd_OE_Y : VitalDelayType01Z;

// Verilog HDL specify path:
(OE *> Y) = (tplh$OE$Y, tphl$OE$Y);
```

- Conditional propagation delay from input port S0 to output port Y.

```
// SDF:
(COND A==0 && B==1 && S1==0
  (IOPATH S0 Y (0.062:0.062:0.062) (0.048:0.048:0.048)
  )
)

-- VHDL generic:
tpd_S0_Y_A_EQ_0_AN_B_EQ_1_AN_S1_EQ_0 :
  VitalDelayType01;

// Verilog HDL specify path:
if ((A == 1'b0) && (B == 1'b1) && (S1 == 1'b0))
  (S0 *> Y) = (tplh$S0$Y, tphl$S0$Y);
```

- Conditional propagation delay from input port A to output port Y.

```
// SDF:
(COND B == 0
  (IOPATH A Y (0.130) (0.098)
  )
)

-- VHDL generic:
tpd_A_Y_B_EQ_0 : VitalDelayType01;
```

```
// Verilog HDL specify path:
if (B == 1'b0)
  (A *> Y) = 0;
```

- Propagation delay from input port CK to output port Q.

```
// SDF:
(IOPATH CK Q (0.100:0.100:0.100) (0.118:0.118:0.118))

-- VHDL generic:
tpd_CK_Q : VitalDelayType01;

// Verilog HDL specify path:
(CK *> Q) = (tplh$CK$Q, tphl$CK$Q);
```

- Conditional propagation delay from input port A to output port Y.

```
// SDF:
(COND B == 1
  (IOPATH A Y (0.062:0.062:0.062) (0.048:0.048:0.048)
  )
)

-- VHDL generic:
tpd_A_Y_B_EQ_1 : VitalDelayType01;

// Verilog HDL specify path:
if (B == 1'b1)
  (A *> Y) = (tplh$A$Y, tphl$A$Y);
```

- Propagation delay from input port CK to output port ECK.

```
// SDF:
(IOPATH CK ECK (0.097:0.097:0.097))

-- VHDL generic:
tpd_CK_ECK : VitalDelayType01;

// Verilog HDL specify path:
(CK *> ECK) = (tplh$CK$ECK, tphl$CK$ECK);
```

- Conditional propagation delay from input port CI to output port S.

```
// SDF:
(COND (A == 0 && B == 0) || (A == 1 && B == 1)
  (IOPATH CI S (0.511) (0.389)
  )
)
```

```
-- VHDL generic:
tpd_CI_S_OP_A_EQ_0_AN_B_EQ_0_CP_OR_OP_A_EQ_1_AN_B_EQ_1_CP:
  VitalDelayType01;
```

```
// Verilog HDL specify path:
if ((A == 1'b0 && B == 1'b0) || (A == 1'b1 && B == 1'b1))
  (CI *> S)  = (tplh$CI$S, tphl$CI$S);
```

- Conditional propagation delay from input port CS to output port S.

```
// SDF:
(COND (A == 1 ^ B == 1 ^ CI1 == 1) &&
    !(A == 1 ^ B == 1 ^ CI0 == 1)
  (IOPATH CS S (0.110) (0.120) (0.120)
              (0.110) (0.119) (0.120)
  )
)
```

```
-- VHDL generic:
tpd_CS_S_OP_A_EQ_1_XOB_B_EQ_1_XOB_CI1_EQ_1_CP_AN_NT_
OP_A_EQ_1_XOB_B_EQ_1_XOB_CI0_EQ_1_CP:
  VitalDelayType01;
```

```
// Verilog HDL specify path:
if ((A == 1'b1 ^ B == 1'b1 ^ CI1N == 1'b0) &&
    !(A == 1'b1 ^ B == 1'b1 ^ CI0N == 1'b0))
  (CS  *> S) = (tplh$CS$S, tphl$CS$S);
```

- Conditional propagation delay from input port A to output port ICO.

```
// SDF:
(COND B == 1 (IOPATH A ICO (0.690)))
```

```
-- VHDL generic:
tpd_A_ICO_B_EQ_1 : VitalDelayType01;
```

```
// Verilog HDL specify path:
if (B == 1'b1)
  (A *> ICO) = (tplh$A$ICO,  tphl$A$ICO);
```

- Conditional propagation delay from input port A to output port CO.

```
// SDF:
(COND (B == 1 ^ C == 1) && (D == 1 ^ ICI == 1)
  (IOPATH A CO (0.263)
  )
)
```

```
-- VHDL generic:
tpd_A_CO_OP_B_EQ_1_XOB_C_EQ_1_CP_AN_OP_D_EQ_1_XOB_ICI
_EQ_1_CP: VitalDelayType01;
```

```
// Verilog HDL specify path:
if ((B == 1'b1 ^ C == 1'b1) && (D == 1'b1 ^ ICI == 1'b1))
  (A *> CO) = (tplh$A$CO,  tphl$A$CO);
```

- Delay from positive edge of CK to Q.

```
// SDF:
(IOPATH (posedge CK) Q (0.410:0.410:0.410)
  (0.290:0.290:0.290))
```

```
-- VHDL generic:
tpd_CK_Q_posedge_noedge : VitalDelayType01;
```

```
// Verilog HDL specify path:
(posedge CK *> Q) = (tplh$CK$Q, tphl$CK$Q);
```

Input Setup Time

- Setup time between posedge of D and posedge of CK.

```
// SDF:
(SETUP (posedge D) (posedge CK) (0.157:0.157:0.157))
```

```
-- VHDL generic:
tsetup_D_CK_posedge_posedge: VitalDelayType;
```

```
// Verilog HDL timing check task:
$setup(posedge CK, posedge D, tsetup$D$CK, notifier);
```

- Setup between negedge of D and posedge of CK.

```
// SDF:
(SETUP (negedge D) (posedge CK) (0.240))

-- VHDL generic:
tsetup_D_CK_negedge_posedge: VitalDelayType;

// Verilog HDL timing check task:
$setup(posedge CK, negedge D, tsetup$D$CK, notifier);
```

- Setup time between posedge of input E with posedge of reference CK.

```
// SDF:
(SETUP (posedge E) (posedge CK) (-0.043:-0.043:-0.043))

-- VHDL generic:
tsetup_E_CK_posedge_posedge : VitalDelayType;

// Verilog HDL timing check task:
$setup(posedge CK, posedge E, tsetup$E$CK, notifier);
```

- Setup time between negedge of input E and posedge of reference CK.

```
// SDF:
(SETUP (negedge E) (posedge CK) (0.101) (0.098))

-- VHDL generic:
tsetup_E_CK_negedge_posedge : VitalDelayType;

// Verilog HDL timing check task:
$setup(posedge CK, negedge E, tsetup$E$CK, notifier);
```

- Conditional setup time between SE and CK.

```
// SDF:
(SETUP (cond E != 1 SE) (posedge CK) (0.155) (0.135))
```

```
-- VHDL generic:
tsetup_SE_CK_E_NE_1_noedge_posedge : VitalDelayType;

// Verilog HDL timing check task:
$setup(posedge CK &&& (E != 1'b1), SE, tsetup$SE$CK,
       notifier);
```

Input Hold Time

- Hold time between posedge of D and posedge of CK.

```
// SDF:
(HOLD (posedge D) (posedge CK) (-0.166:-0.166:-0.166))

-- VHDL generic:
thold_D_CK_posedge_posedge: VitalDelayType;

// Verilog HDL timing check task:
$hold (posedge CK, posedge D, thold$D$CK, notifier);
```

- Hold time between RN and SN.

```
// SDF:
(HOLD (posedge RN) (posedge SN) (-0.261:-0.261:-0.261))

-- VHDL generic:
thold_RN_SN_posedge_posedge: VitalDelayType;

// Verilog HDL timing check task:
$hold (posedge SN, posedge RN, thold$RN$SN, notifier);
```

- Hold time between input port SI and reference port CK.

```
// SDF:
(HOLD (negedge SI) (posedge CK) (-0.110:-0.110:-0.110))

-- VHDL generic:
thold_SI_CK_negedge_posedge: VitalDelayType;

// Verilog HDL timing check task:
$hold (posedge CK, negedge SI, thold$SI$CK, notifier);
```

- Conditional hold time between E and posedge of CK.

```
// SDF:
(HOLD (COND SE ^ RN == 0 E) (posedge CK))
```

```
-- VHDL generic:
thold_E_CK_SE_XOB_RN_EQ_0_noedge_posedge:
  VitalDelayType;
```

```
// Verilog HDL timing check task:
$hold (posedge CK &&& (SE ^ RN == 0), posedge E,
       thold$E$CK, NOTIFIER);
```

Input Setup and Hold Time

- Setup and hold timing check between D and CLK. It is a conditional check. The first delay value is the setup time and the second delay value is the hold time.

```
// SDF:
(SETUPHOLD (COND SE ^ RN == 0 D) (posedge CLK)
  (0.69) (0.32))
```

```
-- VHDL generic (split up into separate setup and hold):
tsetup_D_CK_SE_XOB_RN_EQ_0_noedge_posedge:
  VitalDelayType;
thold_D_CK_SE_XOB_RN_EQ_0_noedge_posedge:
  VitalDelayType;
```

```
-- Verilog HDL timing check (it can either be split up or
-- kept as one construct depending on what appears in the
-- Verilog HDL model):
$setuphold(posedge CK &&& (SE ^ RN == 1'b0)), posedge D,
          tsetup$D$CK,thold$D$CK, notifier);
-- Or as:
$setup(posedge CK &&& (SE ^ RN == 1'b0)), posedge D,
       tsetup$D$CK, notifier);
$hold(posedge CK &&& (SE ^ RN == 1'b0)), posedge D,
      thold$D$CK, notifier);
```

Input Recovery Time

- Recovery time between CLKA and CLKB:

```
// SDF:
(RECOVERY (posedge CLKA) (posedge CLKB)
  (1.119:1.119:1.119))

-- VHDL generic:
trecovery_CLKA_CLKB_posedge_posedge: VitalDelayType;

// Verilog timing check task:
$recovery (posedge CLKB, posedge CLKA,
           trecovery$CLKB$CLKA, notifier);
```

- Conditional recovery time between posedge of CLKA and posedge of CLKB.

```
// SDF:
(RECOVERY (posedge CLKB)
  (COND ENCLKBCLKArec (posedge CLKA)) (0.55:0.55:0.55)
)

-- VHDL generic:
trecovery_CLKB_CLKA_ENCLKBCLKArec_EQ_1_posedge_
posedge: VitalDelayType;

// Verilog timing check task:
$recovery (posedge CLKA && ENCLKBCLKArec, posedge CLKB,
           trecovery$CLKA$CLKB, notifier);
```

- Recovery time between SE and CK.

```
// SDF:
(RECOVERY SE (posedge CK) (1.901))

-- VHDL generic:
trecovery_SE_CK_noedge_posedge: VitalDelayType;

// Verilog timing check task:
$recovery (posedge CK, SE, trecovery$SE$CK, notifier);
```

- Recovery time between RN and CK.

```
// SDF:
(RECOVERY (COND D == 0 (posedge RN)) (posedge CK) (0.8))

-- VHDL generic:
trecovery_RN_CK_D_EQ_0_posedge_posedge:
  VitalDelayType;

// Verilog timing check task:
$recovery (posedge CK && (D == 0), posedge RN,
          trecovery$RN$CK, notifier);
```

Input Removal Time

- Removal time between posedge of E and negedge of CK.

```
// SDF:
(REMOVAL (posedge E) (negedge CK) (0.4:0.4:0.4))

-- VHDL generic:
tremoval_E_CK_posedge_negedge: VitalDelayType;

// Verilog timing check task:
$removal (negedge CK, posedge E, tremoval$E$CK,
          notifier);
```

- Conditional removal time between posedge of CK and SN.

```
// SDF:
(REMOVAL (COND D != 1'b1 SN) (posedge CK) (1.512))

-- VHDL generic:
tremoval_SN_CK_D_NE_1_noedge_posedge : VitalDelayType;

// Verilog timing check task:
$removal (posedge CK &&& (D != 1'b1), SN,
          tremoval$SN$CK, notifier);
```

Period

- Period of input CLKB.

```
// SDF:
(PERIOD CLKB (0.803:0.803:0.803))

-- VHDL generic:
tperiod_CLKB: VitalDelayType;

// Verilog timing check task:
$period (CLKB, tperiod$CLKB);
```

- Period of input port EN.

```
// SDF:
(PERIOD EN (1.002:1.002:1.002))

-- VHDL generic:
tperiod_EN : VitalDelayType;

// Verilog timing check task:
$period (EN, tperiod$EN);
```

- Period of input port TCK.

```
// SDF:
(PERIOD (posedge TCK) (0.220))

-- VHDL generic:
tperiod_TCK_posedge: VitalDelayType;

// Verilog timing check task:
$period (posedge TCK, tperiod$TCK);
```

Pulse Width

- Pulse width of high pulse of CK.

```
// SDF:
(WIDTH (posedge CK) (0.103:0.103:0.103))
```

```
-- VHDL generic:
tpw_CK_posedge: VitalDelayType;

// Verilog timing check task:
$width (posedge CK, tminpwh$CK, 0, notifier);
```

- Pulse width for a low pulse CK.

```
// SDF:
(WIDTH (negedge CK) (0.113:0.113:0.113))

-- VHDL generic:
tpw_CK_negedge: VitalDelayType;

// Verilog timing check task:
$width (negedge CK, tminpwl$CK, 0, notifier);
```

- Pulse width for a high pulse on RN.

```
// SDF:
(WIDTH (posedge RN) (0.122))

-- VHDL generic:
tpw_RN_posedge: VitalDelayType;

// Verilog timing check task:
$width (posedge RN, tminpwh$RN, 0, notifier);
```

Input Skew Time

- Skew between CK and TCK.

```
// SDF:
(SKEW (negedge CK) (posedge TCK) (0.121))

-- VHDL generic:
tskew_CK_TCK_negedge_posedge: VitalDelayType;

// Verilog timing check task:
$skew (posedge TCK, negedge CK, tskew$TCK$CK, notifier);
```

- Skew between SE and negedge of CK.

```
// SDF:
(SKEW SE (negedge CK) (0.386:0.386:0.386))

-- VHDL generic:
tskew_SE_CK_noedge_negedge: VitalDelayType;

// Verilog HDL timing check task:
$skew (negedge CK, SE, tskew$SE$CK, notifier);
```

No-change Setup Time

The SDF NOCHANGE construct maps to both tncsetup and tnchold VHDL generics.

- Nochange setup time between D and negedge CK.

```
// SDF:
(NOCHANGE D (negedge CK) (0.343:0.343:0.343))

-- VHDL generic:
tncsetup_D_CK_noedge_negedge: VitalDelayType;
tnchold_D_CK_noedge_negedge: VitalDelayType;

// Verilog HDL timing check task:
$nochange (negedge CK, D, tnochange$D$CK, notifier);
```

No-change Hold Time

The SDF NOCHANGE construct maps to both tncsetup and tnchold VHDL generics.

- Conditional nochange hold time between E and CLKA.

```
// SDF:
(NOCHANGE (COND RST == 1'b1 (posedge E)) (posedge CLKA)
  (0.312))

-- VHDL generic:
tnchold_E_CLKA_RST_EQ_1_posedge_posedge:
  VitalDelayType;
```

```
tncsetup_E_CLKA_RST_EQ_1_posedge_posedge:
  VitalDelayType;

// Verilog HDL timing check task:
$nochange (posedge CLKA && (RST == 1'b1), posedge E,
            tnochange$E$CLKA, notifier);
```

Port Delay

- Delay to port OE.

  ```
  // SDF:
  (PORT OE (0.266))

  -- VHDL generic:
  tipd_OE: VitalDelayType01;

  // Verilog HDL:
  No explicit Verilog declaration.
  ```

- Delay to port RN.

  ```
  // SDF:
  (PORT RN (0.201:0.205:0.209))

  -- VHDL generic:
  tipd_RN : VitalDelayType01;

  // Verilog HDL:
  No explicit Verilog declaration.
  ```

Net Delay

- Delay on net connected to port CKA.

  ```
  // SDF:
  (NETDELAY CKA (0.134))

  -- VHDL generic:
  tipd_CKA: VitalDelayType01;
  ```

```
// Verilog HDL:
No explicit Verilog declaration.
```

Interconnect Path Delay

- Interconnect path delay from port Y to port D.

```
// SDF:
(INTERCONNECT bcm/credit_manager/U304/Y
  bcm/credit_manager/frame_in/PORT0_DOUT_Q_reg_26_/D
  (0.002:0.002:0.002) (0.002:0.002:0.002))

-- VHDL generic of instance
-- bcm/credit_manager/frame_in/PORT0_DOUT_Q_reg_26_:
tipd_D: VitalDelayType01;
-- The "from" port does not contribute to the timing
-- generic name.

// Verilog HDL:
No explicit Verilog declaration.
```

Device Delay

- Device delay of output SM of instance uP.

```
// SDF:
(INSTANCE uP) . . . (DEVICE SM . . .

-- VHDL generic:
tdevice_uP_SM

// Verilog specify paths:
// All specify paths to output SM.
```

1.5 Complete Syntax

Here is the complete syntax[1] for SDF shown using the BNF form. Terminal names are in uppercase, keywords are in bold uppercase but are case insensitive. The start terminal is delay_file.

absolute_deltype ::= **(ABSOLUTE** del_def { del_def } **)**

alphanumeric ::=
    a | **b** | **c** | **d** | **e** | **f** | **g** | **h** | **i** | **j** | **k** | **l** | **m** | **n** | **o** | **p** |
    q | **r** | **s** | **t** | **u** | **v** | **w** | **x** | **y** | **z**
  | **A** | **B** | **C** | **D** | **E** | **F** | **G** | **H** | **I** | **J** | **K** | **L** | **M** | **N** | **O** | **P** |
    Q | **R** | **S** | **T** | **U** | **V** | **W** | **X** | **Y** | **Z**
  | **_** | **$**
  | decimal_digit

any_character ::=
  character
  | special_character
  | **\"**

arrival_env ::=
  (ARRIVAL [port_edge] port_instance rvalue rvalue
  rvalue rvalue **)**

bidirectskew_timing_check ::=
  (BIDIRECTSKEW port_tchk port_tchk value value **)**

binary_operator ::=
    +
  | **-**
  | *****
  | **/**
  | **%**
  | **==**
  | **!=**
  | **===**
  | **!==**
  | **&&**
  | **||**
  | **<**
  | **<=**
  | **>**
  | **>=**
  | **&**

---

1. The syntax is reprinted with permission from IEEE Std. 1497-2001, Copyright 2001, by IEEE. All rights reserved.

```
    | |
    | ^
    | ^~
    | ~^
    | >>
    | <<
```

bus_net ::= hierarchical_identifier **[** integer **:** integer **]**

bus_port ::= hierarchical_identifier **[** integer **:** integer **]**

ccond ::= **(CCOND** [qstring] timing_check_condition **)**

cell ::= **(CELL** celltype cell_instance { timing_spec } **)**

celltype ::= **(CELLTYPE** qstring **)**

cell_instance ::=
    (INSTANCE [hierarchical_identifier] **)**
    | **(INSTANCE *)**

character ::=
    alphanumeric
    | escaped_character

cns_def ::=
    path_constraint
    | period_constraint
    | sum_constraint
    | diff_constraint
    | skew_constraint

concat_expression ::= **,** simple_expression

condelse_def ::= **(CONDELSE** iopath_def **)**

conditional_port_expr ::=
    simple_expression
    | **(** conditional_port_expr **)**
    | unary_operator **(** conditional_port_expr **)**
    | conditional_port_expr binary_operator
        conditional_port_expr

```
cond_def ::=
  ( COND [ qstring ] conditional_port_expr iopath_def )

constraint_path ::= ( port_instance port_instance )

date ::= ( DATE qstring )

decimal_digit ::= 0 | 1 | 2 | 3 | 4 | 5 | 6 | 7 | 8 | 9

delay_file ::= ( DELAYFILE sdf_header cell { cell } )

deltype ::=
    absolute_deltype
  | increment_deltype
  | pathpulse_deltype
  | pathpulsepercent_deltype

delval ::=
    rvalue
  | ( rvalue rvalue )
  | ( rvalue rvalue rvalue )

delval_list ::=
    delval
  | delval delval
  | delval delval delval
  | delval delval delval delval [ delval ] [ delval ]
  | delval delval delval delval delval delval delval
      [ delval ] [ delval ] [ delval ] [ delval ] [ delval ]

del_def ::=
    iopath_def
  | cond_def
  | condelse_def
  | port_def
  | interconnect_def
  | netdelay_def
  | device_def

del_spec ::= ( DELAY deltype { deltype } )
```

```
departure_env ::=
  ( DEPARTURE [ port_edge ] port_instance rvalue rvalue
    rvalue rvalue )

design_name ::= ( DESIGN qstring )

device_def ::= ( DEVICE [ port_instance ] delval_list )

diff_constraint ::=
  ( DIFF constraint_path constraint_path value [ value ] )

edge_identifier ::=
    posedge
  | negedge
  | 01
  | 10
  | 0z
  | z1
  | 1z
  | z0

edge_list ::=
    pos_pair { pos_pair }
  | neg_pair { neg_pair }

equality_operator ::=
    ==
  | !=
  | ===
  | !==

escaped_character ::=
    \ character
  | \ special_character
  | \"

exception ::= ( EXCEPTION cell_instance { cell_instance } )

hchar := . | /

hierarchical_identifier ::= identifier { hchar identifier }

hierarchy_divider ::= ( DIVIDER hchar )
```

hold_timing_check ::= (**HOLD** port_tchk port_tchk value)

identifier ::= character { character }

increment_deltype ::= (**INCREMENT** del_def { del_def })

input_output_path ::= port_instance port_instance

integer ::= decimal_digit { decimal_digit }

interconnect_def ::=
  (**INTERCONNECT** port_instance port_instance delval_list)

inversion_operator ::=
    !
  | ~

iopath_def ::=
  (**IOPATH** port_spec port_instance { retain_def } delval_list)

lbl_def ::= (identifier delval_list)

lbl_spec ::= (**LABEL** lbl_type { lbl_type })

lbl_type :=
    (**INCREMENT** lbl_def { lbl_def })
  | (**ABSOLUTE** lbl_def { lbl_def })

name ::= (**NAME** qstring)

neg_pair ::=
  (**negedge** signed_real_number [signed_real_number])
  (**posedge** signed_real_number [signed_real_number])

net ::=
    scalar_net
  | bus_net

netdelay_def ::= (**NETDELAY** net_spec delval_list)

net_instance ::=
    net
  | hierarchical_identifier hier_divider_char net

```
net_spec ::=
   port_instance
 | net_instance

nochange_timing_check ::=
  ( NOCHANGE port_tchk port_tchk rvalue rvalue )

pathpulsepercent_deltype ::=
  ( PATHPULSEPERCENT [ input_output_path ] value [ value ] )

pathpulse_deltype ::=
  ( PATHPULSE [ input_output_path ] value [ value ] )

path_constraint ::=
  ( PATHCONSTRAINT [ name ] port_instance port_instance
    { port_instance } rvalue rvalue )

period_constraint ::=
  ( PERIODCONSTRAINT port_instance value [ exception ] )

period_timing_check ::= ( PERIOD port_tchk value )

port ::=
   scalar_port
 | bus_port

port_def ::= ( PORT port_instance delval_list )

port_edge ::= ( edge_identifier port_instance )

port_instance ::=
   port
 | hierarchical_identifier hchar port

port_spec ::=
   port_instance
 | port_edge

port_tchk ::=
   port_spec
 | ( COND [ qstring ] timing_check_condition port_spec )
```

```
pos_pair ::=
  ( posedge signed_real_number [ signed_real_number ] )
  ( negedge signed_real_number [ signed_real_number ] )

process ::= ( PROCESS qstring )

program_name ::= ( PROGRAM qstring )

program_version ::= ( VERSION qstring )

qstring ::= " { any_character } "

real_number ::=
   integer
 | integer [ . integer ]
 | integer [ . integer ] e [ sign ] integer

recovery_timing_check ::=
  ( RECOVERY port_tchk port_tchk value )

recrem_timing_check ::=
   ( RECREM port_tchk port_tchk rvalue rvalue )
 | ( RECREM port_spec port_spec rvalue rvalue
     [ scond ] [ ccond ] )

removal_timing_check ::= ( REMOVAL port_tchk port_tchk value )

retain_def ::= ( RETAIN retval_list )

retval_list ::=
   delval
 | delval delval
 | delval delval delval

rtriple ::=
   signed_real_number : [ signed_real_number ] :
     [ signed_real_number ]
 | [ signed_real_number ] : signed_real_number :
     [ signed_real_number ]
 | [ signed_real_number ] : [ signed_real_number ] :
     signed_real_number
```

```
rvalue ::=
   ( [ signed_real_number ] )
 | ( [ rtriple ] )

scalar_constant ::=
    0
 | 'b0
 | 'B0
 | 1'b0
 | 1'B0
 | 1
 | 'b1
 | 'B1
 | 1'b1
 | 1'B1

scalar_net ::=
   hierarchical_identifier
 | hierarchical_identifier [ integer ]

scalar_node ::=
   scalar_port
 | hierarchical_identifier

scalar_port ::=
   hierarchical_identifier
 | hierarchical_identifier [ integer ]

scond ::= ( SCOND [ qstring ] timing_check_condition )

sdf_header ::=
  sdf_version [ design_name ] [ date ] [ vendor ]
  [ program_name ] [ program_version ] [ hierarchy_divider ]
  [ voltage ] [ process ] [ temperature ] [ time_scale ]

sdf_version ::= ( SDFVERSION qstring )

setuphold_timing_check ::=
   ( SETUPHOLD port_tchk port_tchk rvalue rvalue )
 | ( SETUPHOLD port_spec port_spec rvalue rvalue
     [ scond ] [ ccond ] )

setup_timing_check ::= ( SETUP port_tchk port_tchk value )
```

```
sign ::= + | -

signed_real_number ::= [ sign ] real_number

simple_expression ::=
    ( simple_expression )
  | unary_operator ( simple_expression )
  | port
  | unary_operator port
  | scalar_constant
  | unary_operator scalar_constant
  | simple_expression ? simple_expression : simple_expression
  | { simple_expression [ concat_expression ] }
  | { simple_expression { simple_expression
      [ concat_expression ] } }

skew_constraint ::= ( SKEWCONSTRAINT port_spec value )

skew_timing_check ::= ( SKEW port_tchk port_tchk rvalue )

slack_env ::=
    ( SLACK port_instance rvalue rvalue rvalue rvalue
      [ real_number ] )

special_character ::=
    ! | # | % | & | ' | ( | ) | * | + | , | - | . | / | : | ; | < | = | > | ? |
    @ | [ | \ | ] | ^ | ` | { | | | } | ~

sum_constraint ::=
    ( SUM constraint_path constraint_path { constraint_path }
      rvalue [ rvalue ] )

tchk_def ::=
    setup_timing_check
  | hold_timing_check
  | setuphold_timing_check
  | recovery_timing_check
  | removal_timing_check
  | recrem_timing_check
  | skew_timing_check
  | bidirectskew_timing_check
  | width_timing_check
```

```
    | period_timing_check
    | nochange_timing_check

tc_spec ::= ( TIMINGCHECK tchk_def { tchk_def } )

temperature ::=
    ( TEMPERATURE rtriple )
    | ( TEMPERATURE signed_real_number )

tenv_def ::=
    arrival_env
    | departure_env
    | slack_env
    | waveform_env

te_def ::=
    cns_def
    | tenv_def

te_spec ::= ( TIMINGENV te_def { te_def } )

timescale_number ::= 1 | 10 | 100 | 1.0 | 10.0 | 100.0

timescale_unit ::= s | ms | us | ns | ps | fs

time_scale ::= ( TIMESCALE timescale_number timescale_unit )

timing_check_condition ::=
    scalar_node
    | inversion_operator scalar_node
    | scalar_node equality_operator scalar_constant

timing_spec ::=
    del_spec
    | tc_spec
    | lbl_spec
    | te_spec

triple ::=
    real_number : [ real_number ] : [ real_number ]
    | [ real_number ] : real_number : [ real_number ]
    | [ real_number ] : [ real_number ] : real_number
```

```
unary_operator ::=
    +
  | -
  | !
  | ~
  | &
  | ~&
  | |
  | ~|
  | ^
  | ^~
  | ~^

value ::=
    ( [ real_number ] )
  | ( [ triple ] )

vendor ::= ( VENDOR qstring )

voltage ::=
    ( VOLTAGE rtriple )
  | ( VOLTAGE signed_real_number )

waveform_env ::=
  ( WAVEFORM port_instance real_number edge_list )

width_timing_check ::= ( WIDTH port_tchk value )
```

SPEF

T his chapter describes the Standard Parasitic Extraction Format (SPEF). It is part of the IEEE Std 1481.

2.1 Basics

SPEF allows the description of parasitic information of a design (R, L and C) in an ASCII exchange format. A user can read and check values in a SPEF file, though the user would never create this file manually. It is mainly used to pass parasitic information from one tool to another. Figure 2-1 shows that SPEF can be generated by tools such as a place-and-route tool or a parasitic extraction tool, and then used by a timing analysis tool, in circuit simulation or to perform crosstalk analysis.

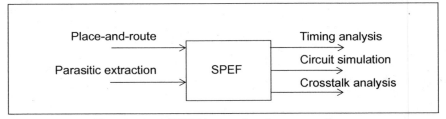

Figure 2-1 SPEF is a tool exchange medium.

Parasitics can be represented at many different levels. SPEF supports the distributed net model, the reduced net model and the lumped capacitance model. In the distributed net model (D_NET), each segment of a net route has its own R and C. In a reduced net model (R_NET), only a single reduced R and C is considered on the load pins of the net and a pie model (C-R-C) is considered on the driver pin of the net. In a lumped capacitance model, only a single capacitance is specified for the entire net. Figure 2-2 shows an example of a physical net route. Figure 2-3 shows the

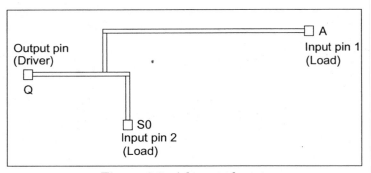

Figure 2-2 A layout of a net.

distributed net model. Figure 2-4 shows the reduced net model and Figure 2-5 shows the lumped capacitance model.

Interconnect parasitics depends on process. SPEF supports the specification of best-case, typical, and worst-case values. Such triplets are allowed for R, L and C values, port slews and loads.

By providing a name map consisting of a map of net names and instance names to indices, the SPEF file size is made effectively smaller, and more importantly, all long names appear in only one place.

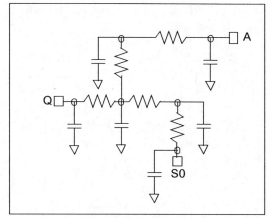

Figure 2-3   Distributed net (D_NET) model.

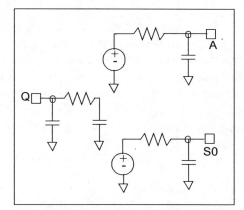

Figure 2-4   Reduced net (R_NET) model.

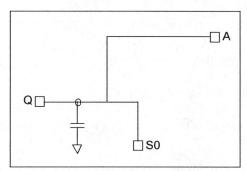

Figure 2-5   Lumped capacitance model.

A SPEF file for a design can be split across multiple files and can also be hierarchical.

2.2   Format

The format of a SPEF file is as follows.

```
header_definition
[ name_map ]
[ power_definition ]
[ external_definition ]
[ define_definition ]
internal_definition
```

The *header definition* contains basic information such as SPEF version number, design name and units for R, L and C. The *name map* specifies the mapping of net names and instance names to indices. The *power definition* declares the power nets and ground nets. The *external definition* defines the ports of the design. The *define definition* identifies instances, whose SPEF is described in additional files. The *internal definition* contains the guts of the file, which are the parasitics of the design.

Figure 2-6 shows an example of a header definition.

*SPEF name

specifies the SPEF version.

*DESIGN name

specifies the design name.

*DATE string

specifies the time stamp when the file was created.

*VENDOR string

specifies the vendor tool that was used to create the SPEF.

```
*SPEF "IEEE 1481-1998"
*DESIGN "ddrphy"
*DATE "Thu Oct 21 00:49:32 2004"
*VENDOR "SGP Design Automation"
*PROGRAM "Galaxy-RCXT"
*VERSION "V2000.06 "
*DESIGN_FLOW "PIN_CAP NONE" "NAME_SCOPE LOCAL"
*DIVIDER /
*DELIMITER :
*BUS_DELIMITER [ ]
*T_UNIT 1.00000 NS
*C_UNIT 1.00000 FF
*R_UNIT 1.00000 OHM
*L_UNIT 1.00000 HENRY

// A comment starts with the two characters "//".

// TCAD_GRD_FILE /cad/13lv/galaxy-rcxt/t013s6ml_fsg.nxtgrd
// TCAD_TIME_STAMP Tue May 14 22:19:36 2002
```

Figure 2-6   A header definition.

```
*PROGRAM string
```

specifies the program that was used to generate the SPEF.

```
*VERSION string
```

specifies the version number of the program that was used to create the SPEF.

```
*DESIGN_FLOW string string string . . .
```

specifies at what stage the SPEF file was created. It describes information about the SPEF file that cannot be derived by reading the file. The pre-defined string values are:

- `EXTERNAL_LOADS` : External loads are fully specified in the SPEF file.

- `EXTERNAL_SLEWS` : External slews are fully specified in the SPEF file.

- **FULL_CONNECTIVITY** : Logical netlist connectivity is present in the SPEF.

- **MISSING_NETS** : Some logical nets may be missing from the SPEF file.

- **NETLIST_TYPE_VERILOG** : Uses Verilog HDL type naming conventions.

- **NETLIST_TYPE_VHDL87** : Uses VHDL87 naming convention.

- **NETLIST_TYPE_VHDL93** : Uses VHDL93 netlist naming convention.

- **NETLIST_TYPE_EDIF** : Uses EDIF type naming convention.

- **ROUTING_CONFIDENCE** positive_integer : Default routing confidence number for all nets, basically the level of accuracy of the parasitics.

- **ROUTING_CONFIDENCE_ENTRY** positive_integer string : Supplements the routing confidence values.

- **NAME_SCOPE LOCAL | FLAT** : Specifies whether paths in the SPEF file are relative to file or to top of design.

- **SLEW_THRESHOLDS** low_input_threshold_percent high_input_threshold_percent : Specifies the default input slew threshold for the design.

- **PIN_CAP NONE | INPUT_OUTPUT | INPUT_ONLY** : Specifies what type of pin capacitances are included as part of total capacitance. The default is **INPUT_OUTPUT**.

The line in the header definition:

    *DIVIDER /

specifies the hierarchy delimiter. Other characters that can be used are ., :, and /.

    *DELIMITER :

specifies the delimiter between an instance and its pin. Other possible characters that can be used are ., /, :, or |.

    *BUS_DELIMITER [ ]

specifies the prefix and suffix that are used to identify a bit of a bus. Other possible characters that can be used for prefix and suffix are {, (, <, :, . and },), >.

```
*T_UNIT positive_integer NS | PS
```

specifies the time unit.

```
*C_UNIT positive_integer PF | FF
```

specifies the capacitance unit.

```
*R_UNIT positive_integer OHM | KOHM
```

specifies the resistance unit.

```
*L_UNIT positive_integer HENRY | MH | UH
```

specifies the inductance unit.

A *comment* in a SPEF file can appear in two forms.

```
// Comment - until end of line.

/* This comment can
extend across multiple
lines */
```

Figure 2-7 shows an example of a *name map*. It is of the form:

```
*NAME_MAP
*positive_integer name
*positive_integer name
 . . .
```

The name map specifies the mapping of names to unique integer values (their indices). The name map helps in reducing the file size by making all future references of the name by the index. A name can be a net name or an instance name. Given the name map in Figure 2-7, the names can later be referenced in the SPEF file by using their index, such as:

```
*NAME_MAP
*1 memclk
*2 memclk_2x
*3 reset_
*4 refresh
*5 resync
*6 int_d_out[63]
*7 int_d_out[62]
*8 int_d_out[61]
*9 int_d_out[60]
*10 int_d_out[59]
*11 int_d_out[58]
*12 int_d_out[57]

. . .
*364 mcdll_write_data/write19/d_out_2x_reg_19
*366 mcdll_write_data/write20/d_out_2x_reg_20
*368 mcdll_write_data/write21/d_out_2x_reg_21

. . .
*5423 mcdll_read_data/read21/capture_data[53]

. . .
*5426 mcdll_read_data/read21/capture_pos_0[21]

. . .
*11172 Tie_VSSQ_assign_buf_318_N_1

. . .
*14954 test_se_15_S0
*14955 wr_sdly_course_enc[0]_L0
*14956 wr_sdly_course_enc[0]_L0_1
*14957 wr_sdly_course_enc[0]_S0
```

Figure 2-7 A name map.

```
*364:D          // D pin of instance
        // mcdll_write_data/write19/d_out_2x_reg_19
*11172:Y        // Y pin of instance
                // Tie_VSSQ_assign_buf_318_N_1
*5426:116       // Internal node of net
        // mcdll_read_data/read21/capture_pos_0[21]
*5426:10278     // Internal node of net *5426
*12             // The net int_d_out[57]
```

The name map thus avoids repeating long names and their paths by using their unique integer representation.

The *power definition* section defines the power and ground nets.

```
*POWER_NETS net_name net_name . . .
*GROUND_NETS net_name net_name . . .
```

Here are some examples.

```
*POWER_NETS VDDQ
*GROUND_NETS VSSQ
```

The *external definition* contains the definition of the logical and physical ports of the design. Figure 2-8 shows an example of logical ports. Logical ports are described in the form:

```
*PORTS
port_name direction { conn_attribute }
port_name direction { conn_attribute }
. . .
```

where a *port_name* can be the port index of form `*positive_integer`. The *direction* is I for input, O for output and B for bidirectional. *Connection attributes* are optional, and can be the following:

- ***C** `number  number` : Coordinates of the port.
- ***L** `par_value` : Capacitive load of the port.
- ***S** `par_value  par_value` : Defines the shape of the waveform on the port.
- ***D** `cell_type` : Defines the driving cell of the port.

Physical ports in a SPEF file are defined using:

```
*PHYSICAL_PORTS
pport_name direction { conn_attribute }
pport_name direction { conn_attribute }
. . .
```

The *define definition* section defines entity instances that are referenced in the current SPEF file but whose parasitics are described in additional SPEF files.

```
*PORTS
*1 I
*2 I
*3 I
*4 I
*5 I
*6 I
*7 I
*8 I
*9 I
*10 I
*11 I
. . .
*450 O
*451 O
*452 O
*453 O
*454 O
*455 O
*456 O
```

Figure 2-8   An external definition.

```
*DEFINE instance_name { instance_name } entity_name
*PDEFINE physical_instance entity_name
```

The *PDEFINE is used when the entity instance is a physical partition (instead of a logical hierarchy). Here are some examples.

```
*DEFINE core/u1ddrphy core/u2ddrphy "ddrphy"
```

This implies that there would be another SPEF file with a *DESIGN value of ddrphy - this file would contain the parasitics for the design ddrphy. It is possible to have physical and logical hierarchy. Any nets that cross the hierarchical boundaries have to be described as distributed nets (D_NET).

The *internal definition* forms the guts of the SPEF file - it describes the parasitics for the nets in the design. There are basically two forms: the *distributed net*, D_NET, and the *reduced net*, R_NET. Figure 2-9 shows an example of a distributed net definition.

```
*D_NET *5426 0.899466

*CONN
*I *14212:D I *C 21.7150 79.2300
*I *14214:Q O *C 21.4950 76.6000 *D DFFQX1

*CAP
1 *5426:10278 *5290:8775 0.217446
2 *5426:10278 *16:3754 0.0105401
3 *5426:10278 *5266:9481 0.0278254
4 *5426:10278 *5116:9922 0.113918
5 *5426:10278 0.529736

*RES
1 *5426:10278 *14212:D 0.340000
2 *5426:10278 *5426:10142 0.916273
3 *5426:10142 *14214:Q 0.340000
*END
```

Figure 2-9 Distributed net parasitics for net *5426.

In the first line,

```
*D_NET *5426 0.899466
```

*5426 is the net index (see name map for the net name) and 0.899466 is the total capacitance value on the net. The capacitance value is the sum of all capacitances on the net including cross-coupling capacitances that are assumed to be grounded, and including load capacitances. It may or may not include pin capacitances depending on the setting of PIN_CAP in the *DESIGN_FLOW definition.

The *connectivity section* describes the drivers and loads for the net. In:

```
*CONN
*I *14212:D I *C 21.7150 79.2300
*I *14214:Q O *C 21.4950 76.6000 *D DFFQX1
```

*I refers to an internal pin (*P is used for a port), *14212:D refers to the D pin of instance *14212 which is an index (see name map for actual name). "I" says that it is a load (input pin) on the net. "O" says that it is a driver (output pin) on the net. *C and *D are as defined earlier in connection attributes - *C defines the coordinates of the pin and *D defines the driving cell of the pin.

The *capacitance section* describes the capacitances of the distributed net. The capacitance unit is as specified earlier with *C_UNIT.

```
*CAP
1 *5426:10278 *5290:8775 0.217446
2 *5426:10278 *16:3754 0.0105401
3 *5426:10278 *5266:9481 0.0278254
4 *5426:10278 *5116:9922 0.113918
5 *5426:10278 0.529736
```

The first number is the capacitance identifier. There are two forms of capacitance specification, first through fourth are of one form and the fifth is of the second form. The first form (first through fourth) specifies the cross-coupling capacitances between two nets, while the second form (with id 5) specifies the capacitance to ground. So in capacitance id 1, the cross-coupling capacitance between nets *5426 and *5290 is 0.217446. And in capacitance id 5, the capacitance to ground is 0.529736. Notice that the first node name is necessarily the net name for the D_NET that is being described. The positive integer following the net index (10278 in *5426:10278) specifies an internal node or junction point. So capacitance id 4 states that there is a coupling capacitance between net *5426 with internal node 10278 and net *5116 with internal node 9922 and the value of this coupling capacitance is 0.113918.

The *resistance section* describes the resistances of the distributed net. The resistance unit is as specified with *R_UNIT.

```
*RES
1 *5426:10278 *14212:D 0.340000
2 *5426:10278 *5426:10142 0.916273
3 *5426:10142 *14214:Q 0.340000
```

The first field is the resistance identifier. So there are three resistance components for this net. The first one is between the internal node *5426:10278 to the D pin on *14212 and the resistance value is 0.34. The capacitance and resistance section can be better understood with the RC network shown pictorially in Figure 2-10.

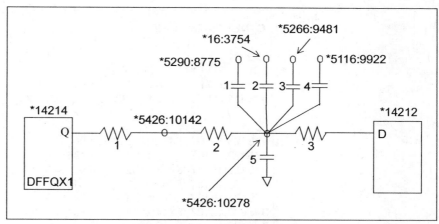

Figure 2-10   RC for net *5426.

Figure 2-11 shows another example of a distributed net. This net has one driver and two loads and the total capacitance on the net is 2.69358. Figure 2-12 shows the RC network that corresponds to the distributed net specification.

In general, an internal definition can comprise of the following specifications:

- **D_NET**: Distributed RC network form of a logical net.
- **R_NET**: Reduced RC network form of a logical net.
- **D_PNET**: Distributed form of a physical net.
- **R_PNET**: Reduced form of a physical net.

Here is the syntax.

```
*D_NET net_index total_cap [*V routing_confidence ]
 [ conn_section ]
 [ cap_section ]
 [ res_section ]
 [ inductance_section ]
*END

*R_NET net_index total_cap [ *V routing_confidence ]
 [ driver_reduction ]
*END
```

```
*D_NET *5423 2.69358

*CONN
*I *14207:D I *C 21.7450 94.3150
*I *14205:D I *C 21.7450 90.4900
*I *14211:Q O *C 21.4900 83.8800 *D DFFQX1

*CAP
1 *5423:10107 *547:12722 0.202686
2 *5423:10107 *5116:10594 0.104195
3 *5423:10107 *5233:9552 0.208867
4 *5423:10107 *5265:9483 0.0225810
5 *5423:10107 *267:9668 0.0443454
6 *5423:10107 *5314:7853 0.120589
7 *5423:10212 *2109:996 0.0293744
8 *5423:10212 *5187:7411 0.526945
9 *5423:14640 *6577:10075 0.126929
10 *5423:10213 1.30707

*RES
1 *5423:10107 *5423:10212 2.07195
2 *5423:10107 *5423:10106 0.340000
3 *5423:10212 *5423:10211 0.340000
4 *5423:10212 *5423:14640 1.17257
5 *5423:14640 *5423:10213 0.340000
6 *5423:10213 *14207:D 0.0806953
7 *5423:10211 *14205:D 0.210835
8 *5423:10106 *14211:Q 0.0932139
*END
```

Figure 2-11   Another example of a distributed net *5423.

```
*D_PNET pnet_index total_cap [*V routing_confidence ]
   [ pconn_section ]
   [ pcap_section ]
   [ pres_section ]
   [ pinduc_section ]
*END

*R_PNET pnet_index total_cap [*V routing_confidence ]
   [ pdriver_reduction ]
*END
```

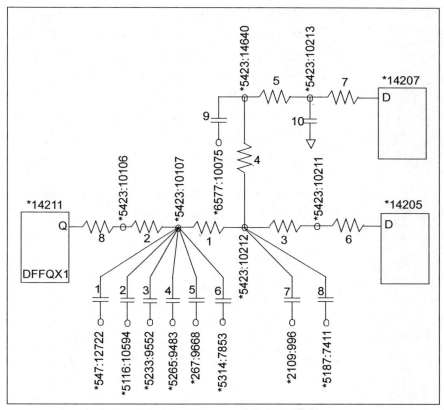

Figure 2-12   RC network for D_NET *5423.

The *inductance section* is used to specify inductances and the format is similar to the resistance section. The ***v** is used to specify the accuracy of the parasitics of the net. These can be specified individually with a net or can be specified globally using the ***DESIGN_FLOW** statement with the ROUTING_CONFIDENCE value, such as:

*DESIGN_FLOW "ROUTING_CONFIDENCE 100"

which specifies that the parasitics were extracted after final cell placement and final route and 3d extraction was used. Other possible values of routing confidence are:

- 10: Statistical wireload model
- 20: Physical wireload model

- 30: Physical partitions with locations, and no cell placement
- 40: Estimated cell placement with steiner tree based route
- 50: Estimated cell placement with global route
- 60: Final cell placement with steiner route
- 70: Final cell placement with global route
- 80: Final cell placement, final route, 2d extraction
- 90: Final cell placement, final route, 2.5d extraction
- 100: Final cell placement, final route, 3d extraction

A *reduced net* is a net that has been reduced from a distributed net form. There is one driver reduction section for each driver on a net. The driver reduction section is of the form:

```
*DRIVER pin_name
*CELL cell_type
// Driver reduction: one such section for each driver
// of net:
*C2_R1_C1 cap_value res_value cap_value
*LOADS      // One following set for each load on net:
*RC pin_name rc_value
*RC pin_name rc_value
. . .
```

The *C2_R1_C1 shows the parasitics for the pie model on the driver pin of the net. The *rc_value* in *RC construct is the Elmore delay (R*C). Figure 2-13 shows an example of a reduced net SPEF and Figure 2-14 shows

```
*R_NET *1200 2.995
*DRIVER *1201:Q
*CELL SEDFFX1
*C2_R1_C1 0.511 2.922 0.106
*LOADS
*RC *1202:A 1.135
*RC *1203:A 0.946
*END
```

Figure 2-13   Reduced net example.

the RC network pictorially.

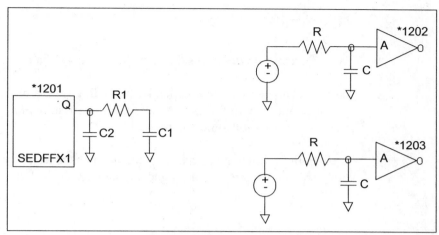

Figure 2-14 Reduced net model.

A *lumped capacitance* model is described using either a *D_NET or a
*R_NET construct with just the total capacitance and with no other infor-
mation. Here are examples of lumped capacitance declarations.

```
*D_NET *1 80.2096
*CONN
*I *2:Y O *L 0 *D CLKMX2X2
*P *1 O *L 0
*END

*R_NET *17 58.5204
*END
```

Values in a SPEF file can be in a triplet form that represents the process variations, such as:

```
0.243:0.269:0.300
```

0.243 is the best-case value, 0.269 is the typical value and 0.300 is the
worst-case value.

2.3   Complete Syntax

This section describes the complete syntax[1] of a SPEF file.

A character can be escaped by preceding with a backslash (\). Comments come in two forms: // starts a comment until end of line, while /* . . . */ is a multi-line comment.

In the following syntax, bold characters such as (, [are part of the syntax. All constructs are arranged alphabetically and the start symbol is SPEF_file.

```
alpha ::= upper | lower

bit_identifier ::=
    identifier
  | <identifier><prefix_bus_delim><digit>{<digit>}
    [ <suffix_bus_delim> ]

bus_delim_def ::=
    *BUS_DELIMITER prefix_bus_delim [ suffix_bus_delim ]

cap_elem ::=
    cap_id node_name par_value
  | cap_id node_name node_name2 par_value

cap_id ::= pos_integer

cap_load ::= *L par_value

cap_scale ::= *C_UNIT pos_number cap_unit

cap_sec ::= *CAP cap_elem { cap_elem }

cap_unit ::= PF | FF

cell_type ::= index | name
```

---

1. Syntax is reprinted here with permission from IEEE Std. 1481-1999, Copyright 1999, by IEEE. All rights reserved.

cnumber ::= (real_component imaginary_component)

complex_par_value ::=
   cnumber
 | number
 | cnumber:cnumber:cnumber
 | number:number:number

conf ::= pos_integer

conn_attr ::= coordinates | cap_load | slews | driving_cell

conn_def ::=
  *P external_connection direction { conn_attr }
| *I internal_connection direction { conn_attr }

conn_sec ::=
  *CONN conn_def { conn_def } { internal_node_coord }

coordinates ::= *C number number

date ::= *DATE qstring

decimal ::= [sign]<digit>{<digit>}.{<digit>}

define_def ::= define_entry { define_entry }

define_entry ::=
   *DEFINE inst_name { inst_name } entity
 | *PDEFINE physical_inst entity

design_flow ::= *DESIGN_FLOW qstring [qstring]

design_name ::= *DESIGN qstring

digit ::= 0 - 9

direction ::= I | B | O

driver_cell ::= *CELL cell_type

driver_pair ::= *DRIVER pin_name

driver_reduc ::= driver_pair driver_cell pie_model load_desc

driving_cell ::= *D cell_type

d_net ::=
  *D_NET net_ref total_cap [routing_conf]
  [conn_sec]
  [cap_sec]
  [res_sec]
  [induc_sec]
  *END

d_pnet ::=
  *D_PNET pnet_ref total_cap [routing_conf]
  [pconn_sec]
  [pcap_sec]
  [pres_sec]
  [pinduc_sec]
  *END

entity ::= qstring

escaped_char ::= \<escaped_char_set>

escaped_char_set ::= <special_char> | "

exp ::= <radix><exp_char><integer>

exp_char ::= E | e

external_connection ::= port_name | pport_name

external_def ::=
  port_def [physical_port_def]
 | physical_port_def

float ::=
  decimal
 | fraction
 | exp

fraction ::= [sign].<digit>{<digit>}

```
ground_net_def ::= *GROUND_NETS net_name { net_name }

hchar ::= . | / | : | |

header_def ::=
  SPEF_version
  design_name
  date
  vendor
  program_name
  program_version
  design_flow
  hierarchy_div_def
  pin_delim_def
  bus_delim_def
  unit_def

hierarchy_div_def ::= *DIVIDER hier_delim

hier_delim ::= hchar

identifier ::= <identifier_char>{<identifier_char>}

identifier_char ::=
    <escaped_char>
  | <alpha>
  | <digit>
  | _

imaginary_component ::= number

index ::= *<pos_integer>

induc_elem ::= induc_id node_name node_name par_value

induc_id ::= pos_integer

induc_scale ::= *L_UNIT pos_number induc_unit

induc_sec ::= *INDUC induc_elem { induc_elem }

induc_unit ::= HENRY | MH | UH
```

inst_name ::= index | path

integer ::= [sign]<digit>{<digit>}

internal_connection ::= pin_name | pnode_ref

internal_def ::= nets { nets }

internal_node_coord ::= *N internal_node_name coordinates

internal_node_name ::= <net_ref><pin_delim><pos_integer>

internal_pnode_coord ::= *N internal_pnode_name coordinates

internal_pnode_name ::= <pnet_ref><pin_delim><pos_integer>

load_desc ::= *LOADS rc_desc { rc_desc }

lower ::= a - z

mapped_item ::=
    identifier
  | bit_identifier
  | path
  | name
  | physical_ref

name ::= qstring | identifier

name_map ::= *NAME_MAP name_map_entry { name_map_entry }

name_map_entry ::= index mapped_item

neg_sign ::= -

nets ::= d_net | r_net | d_pnet | r_pnet

net_name ::= net_ref | pnet_ref

net_ref ::= index | path

net_ref2 ::= net_ref

```
node_name ::=
    external_connection
  | internal_connection
  | internal_node_name
  | pnode_ref

node_name2 ::=
    node_name
  | <pnet_ref><pin_delim><pos_integer>
  | <net_ref2><pin_delim><pos_integer>

number ::= integer | float

partial_path ::= <hier_delim><bit_identifier>

partial_physical_ref ::= <hier_delim><physical_name>

par_value ::= float | <float>:<float>:<float>

path ::=
  [<hier_delim>]<bit_identifier>{<partial_path>}
  [<hier_delim>]

pcap_elem ::=
    cap_id pnode_name par_value
  | cap_id pnode_name pnode_name2 par_value

pcap_sec ::= *CAP pcap_elem { pcap_elem }

pconn_def ::=
    *P pexternal_connection direction { conn_attr }
  | *I internal_connection direction { conn_attr }

pconn_sec ::=
  *CONN pconn_def { pconn_def } { internal_pnode_coord }

pdriver_pair ::= *DRIVER internal_connection

pdriver_reduc ::= pdriver_pair driver_cell pie_model load_desc

pexternal_connection ::= pport_name
```

```
physical_inst ::= index | physical_ref

physical_name ::= name

physical_port_def ::=
  *PHYSICAL_PORTS pport_entry { pport_entry }

physical_ref ::= <physical_name>{<partial_physical_ref>}

pie_model ::=
  *C2_R1_C1 par_value par_value par_value

pin ::= index | bit_identifier

pinduc_elem ::= induc_id pnode_name pnode_name par_value

pinduc_sec ::=
  *INDUC
  pinduc_elem
  { pinduc_elem }

pin_delim ::= hchar

pin_delim_def ::= *DELIMITER pin_delim

pin_name ::= <inst_name><pin_delim><pin>

pnet_ref ::= index | physical_ref

pnet_ref2 ::= pnet_ref

pnode ::= index | name

pnode_name ::=
    pexternal_connection
  | internal_connection
  | internal_pnode_name
  | pnode_ref

pnode_name2 ::=
    pnode_name
  | <net_ref><pin_delim><pos_integer>
  | <pnet_ref2><pin_delim><pos_integer>
```

```
pnode_ref ::= <physical_inst><pin_delim><pnode>

pole ::= complex_par_value

pole_desc ::= *Q pos_integer pole { pole }

pole_residue_desc ::= pole_desc residue_desc

port_def ::=
  *PORTS
  port_entry
  { port_entry }

pos_decimal ::= <digit>{<digit>}.{<digit>}

port ::= index | bit_identifier

port_entry ::= port_name direction { conn_attr }

port_name ::= [<inst_name><pin_delim>]<port>

pos_exp ::= pos_radix exp_char integer

pos_float ::= pos_decimal | pos_fraction | pos_exp

pos_fraction ::= .<digit>{<digit>}

pos_integer ::= <digit>{<digit>}

pos_number ::= pos_integer | pos_float

pos_radix ::= pos_integer | pos_decimal | pos_fraction

pos_sign ::= +

power_def ::=
    power_net_def [ ground_net_def ]
  | ground_net_def

power_net_def ::= *POWER_NETS net_name { net_name }

pport ::= index | name
```

```
pport_entry ::= pport_name direction { conn_attr }

pport_name ::= [<physical_inst><pin_delim>]<pport>

prefix_bus_delim ::= { | [ | ( | < | : | .

pres_elem ::= res_id pnode_name pnode_name par_value

pres_sec ::=
  *RES
  pres_elem
  { pres_elem }

program_name ::= *PROGRAM qstring

program_version ::= *VERSION qstring

qstring ::= "{qstring_char}"

qstring_char ::= special_char | alpha | digit | white_space | _

radix ::= decimal | fraction

rc_desc ::= *RC pin_name par_value [ pole_residue_desc ]

real_component ::= number

residue ::= complex_par_value

residue_desc := *K pos_integer residue { residue }

res_elem ::= res_id node_name node_name par_value

res_id ::= pos_integer

res_scale ::= *R_UNIT pos_number res_unit

res_sec ::=
  *RES
  res_elem
  { res_elem }

res_unit ::= OHM | KOHM
```

```
routing_conf ::= *V conf

r_net ::=
  *R_NET net_ref total_cap [ routing_conf ]
  { driver_reduc }
  *END

r_pnet ::=
  *R_PNET pnet_ref total_cap [ routing_conf ]
  { pdriver_reduc }
  *END

sign ::= pos_sign | neg_sign

slews ::= *S par_value par_value [ threshold threshold ]

special_char ::=
    ! | # | $ | % | & | ` | ( | ) | * | + | , | - | . | / | : | ; | < | = | >
  | ? | @ | [ | \ | ] | ^ | ' | { | | | } | ~

SPEF_file ::=
  header_def
  [ name_map ]
  [ power_def ]
  [ external_def ]
  [ define_def ]
  internal_def

SPEF_version ::= *SPEF qstring

suffix_bus_delim ::= ] | } | ) | >

threshold ::=
    pos_fraction
  | <pos_fraction>:<pos_fraction>:<pos_fraction>

time_scale ::= *T_UNIT pos_number time_unit

time_unit ::= NS | PS

total_cap ::= par_value

unit_def ::= time_scale cap_scale res_scale induc_scale
```

upper ::= **A** - **Z**

vendor ::= *****VENDOR** qstring

white_space ::= space | tab

❑

LEF

T his chapter describes the Library Exchange Format (LEF). It is an industry defacto standard.

3.1   Basics

LEF supports the physical description of one or more macrocell definitions. It describes the physical model, as a black-box, of a design or a macrocell in an ASCII form. A LEF file defines a library of macrocell definitions.

A macrocell definition in a LEF file is used during layout to provide the physical dimensions of a macro cell being placed. The macrocell definition is typically generated by a layout tool for the design, such that the macrocell can be placed or used in another larger design.

A macrocell definition in a LEF file is like a component declaration in VHDL. A macrocell definition is later used (instantiated) in a Design Exchange Format (DEF) file - the topic of the next chapter.

A LEF file can also be used by a parasitic extraction tool to obtain physical information of macrocells being used in a design. Figure 3-1 shows a typical usage of a LEF file.

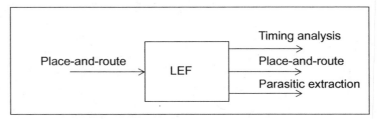

Figure 3-1   LEF is a tool exchange medium.

A library vendor or an IP vendor typically provide LEF files for the macrocells that the vendor supports for use in physical design.

3.2   Format

Each line in a LEF file is limited to 2048 characters. A basic statement in a LEF file ends with a space followed by a semicolon (;). The length unit is 1 micron. The format of a LEF file is as follows.

```
[ VERSION statement ]
[ BUSBITCHARS statement ]
[ DIVIDERCHAR statement ]
[ UNITS statement ]
[ MANUFACTURINGGRID statement ]
[ USEMINSPACING statement ]
[ CLEARANCEMEASURE statement ]
[ PROPERTYDEFINITIONS statement ]
[ LAYER (Nonrouting) statement
| LAYER (Routing) statement ] ...
[ SPACING statement ]
[ MAXVIASTACK statement ]
[ VIA statement ] ...
[ VIARULE statement ] ...
```

```
[ VIARULE GENERATE statement ] ...
[ NONDEFAULTRULE statement ] ...
[ SITE statement ] ...
[ MACRO statement
  [ PIN statement ] ...
  [ OBS statement ... ] ] ...
[ BEGINEXT statement ] ...
[ END LIBRARY ]
```

A LEF file consists of a header followed by one or more macro definitions. The header contains basic information such as version, case sensitivity, hierarchy character, bit specifier and site information. Each macro definition defines the macro, its size and its pin definitions and the physical characteristics of each of the macro's pins. Statements in a LEF file can be in any order but with the only rule that data must be defined prior to its use. For example, a via must be defined before it is referenced in other statements.

Figure 3-2 shows an example of a header definition.[1]

```
VERSION 5.3 ;
NAMESCASESENSITIVE ON ;
DIVIDERCHAR "/" ;
BUSBITCHARS "[ ]" ;

UNITS
    DATABASE MICRONS 2000 ;
END UNITS

########## SITE ###############
SITE rf1_16x24cm2_site
  CLASS CORE ;
  SIZE 109.095 BY 94.205 ;
END rf1_16x24cm2_site

SITE rf1_256x32cm4sw8_site
  CLASS CORE ;
  SIZE 224.355 BY 178.805 ;
END rf1_256x32cm4sw8_site
. . .
```

Figure 3-2   A header section.

---

1. The examples shown are based on LEF versions 5.3 and 5.4, while the grammar and syntax shown are for version 5.6.

```
VERSION 5.3 ;
```

specifies the LEF version.

```
NAMESCASESENSITIVE ON ;
```

specifies that the names are case sensitive.

```
DIVIDERCHAR "/" ;
```

specifies the character used in specifying hierarchy in LEF names. The backslash character "\" can be used to escape the hierarchy character if it appears in a LEF name. The default divider character is the "/" character.

The BUSBITCHARS statement specifies how bus bits are shown in a LEF file, for example, as HADDR[1]. An example of a BUSBITCHARS statement is:

```
BUSBITCHARS "[]" ;
```

If the bus bit character is part of the name, then the backslash character "\" can be used to escape the bus bit character. For example, in the name H\[addr\][31], the first "[ ]" characters are part of the name and are not the bus bit characters. In the absence of a BUSBITCHARS statement in a LEF file, the default bus bit characters are "[ ]".

The units section,

```
UNITS
   DATABASE MICRONS 2000;
END UNITS
```

specifies the factor by which the length numbers that appear in a LEF file are multiplied by when a LEF file is read by a LEF reader, that is, it represents the precision of the length values. Other units that can be specified are:

```
TIME NANOSECONDS factor ;
CAPACITANCE PICOFARADS factor ;
RESISTANCE OHMS factor ;
POWER MILLIWATTS factor ;
```

```
CURRENT MILLIAMPS factor ;
VOLTAGE VOLTS factor ;
FREQUENCY MEGAHERTZ factor ;
```

If the `DATABASE MICRONS` is not specified, a default factor of 100 is used.

The # character is used to specify the start of a comment.

########## SITE ###############

A LEF file can have one or more site specifications. A site specification defines a placement site in the design. In the site specification:

```
SITE rf1_16x24cm2_site
  CLASS CORE ;
  SIZE 109.095 BY 94.205 ;
END rf1_16x24cm2_site
```

the placement information is specified for a memory whose placement site name is `rf1_16x24cm2_site`. It is of class `CORE` (the other valid value is a `PAD` which would be applicable for an `IO` pad). The class specification is followed by the size specification of the macro - each number is expressed in microns. A site specification may optionally contain a `SYMMETRY` construct that can take values `X`, `Y` or `R90`; the value indicates the allowable orientations of the macro.

Figure 3-3 shows an example of one macro in the LEF file. A LEF file can contain one or more such macro definitions.

The macro definition begins with the macro name `AND2X1` followed by the type of the macro, which is `CORE`. Additional `CLASS` types are `COVER`, `RING`, `BLOCK`, `PAD`, and `ENDCAP`. If no `CLASS` type is specified, the default is `CORE`.

The `FOREIGN` specification specifies the GDSII (foreign) structure name that is to be used. The coordinates "`0.000  0.000`" specify the macro origin as offset from the foreign origin. The orientation `N` specifies the orientation of the foreign cell when macro is in north orientation.

```
MACRO AND2X1
  CLASS CORE ;
  FOREIGN AND2X1 0.000 0.000 N ;
  SIZE 1.840 BY 3.690 ;
  ORIGIN 0.000 0.000 ;
  SYMMETRY X Y ;

  PIN A
    DIRECTION INPUT ;
    USE SIGNAL ;
    PORT
      LAYER METAL1 ;
      RECT 0.125 1.400 0.405 1.990 ;
    END
  END A

  PIN B
    DIRECTION INPUT ;
    USE SIGNAL ;
    PORT
      LAYER METAL1 ;
      RECT 0.585 1.290 0.795 2.145 ;
      RECT 0.585 1.885 0.845 2.145 ;
    END
  END B

  PIN Y
    DIRECTION OUTPUT ;
    USE SIGNAL ;
    PORT
      LAYER METAL1 ;
      RECT 1.365 2.300 1.625 2.900 ;
      RECT 1.505 0.955 1.715 2.585 ;
    END
  END Y

  PIN VSS
    DIRECTION INOUT ;
    USE GROUND ;
    PORT
      LAYER METAL1 ;
      RECT 0.000 -0.250 1.840 0.250 ;
      RECT 0.935 -0.250 1.195 0.405 ;
    END
  END VSS

  PIN VDD
    DIRECTION INOUT ;
    USE POWER ;
    PORT
      LAYER METAL1 ;
      RECT 0.000 3.440 1.840 3.940 ;
    END
  END VDD

  OBS
    LAYER METAL1 ;
    RECT 1.185 1.710 1.325 1.970 ;
    RECT 0.225 0.535 0.385 1.110 ;
    RECT 0.315 2.325 0.575 2.585 ;
  END
END AND2X1
```

Figure 3-3 A macro definition in a LEF file.

The SIZE specification "1.840 BY 3.690" specifies the minimum bounding rectangle of the macro, which includes all pins and blockages.

The ORIGIN specification specifies the origin of the macro. It refers to the lower left corner of the macro. All other coordinates are specified with respect to this origin.

There are a number of PIN specifications in the macro definition. For each pin, its name is given followed by whether it is an INPUT, OUTPUT, INOUT or a FEEDTHRU. This is followed by a USE specification that specifies the functionality of the pin which is one of SIGNAL, ANALOG, POWER, GROUND or CLOCK.

A PIN specification contains a port statement starting with PORT. This statement defines a set of geometries that are electrically equivalent points. The port statement for pin A specifies that the port is on layer 1 and that the rectangle that comprises the pin in the physical layout is "0.125 1.400 0.405 1.990". The RECT statement specifies x1, y1 to x2, y2. The two points are the opposite corners of the rectangle.

The last section in the macro definition describes the set of obstructions (blockages) in the macro.

Figure 3-4 shows another example of a LEF file. The header information provides the LEF version number, the case sensitivity of names, the hierarchy divider separator character, the bus bit characters and the length multiplication factor of 2000.

The SITE specification describes the placement site in the design. It is of type CORE and the size of the placement site is specified to be 1822.00u[1] by 1123.00u. This LEF library contains a description of only one macro, that of the ARM926EJS. The macro's size, origin and location as it relates to the placement site are specified next, and the macro's allowable orientations are set to be X, Y (N and FN are other legal values) and R90.

---

1. "u" indicates "micron".

```
VERSION 5.3 ;
NAMESCASESENSITIVE ON ;
DIVIDERCHAR "/" ;
BUSBITCHARS "[ ]" ;

UNITS
  DATABASE MICRONS 2000 ;
END UNITS

########## SITE ###############
SITE ARM926EJS_site
  CLASS CORE ;
  SIZE 1822.000 BY 1123.000 ;
END ARM926EJS_site
########## MACROS ##############

MACRO ARM926EJS
  CLASS BLOCK ;
  SIZE 1822.000 BY 1123.000 ;
  ORIGIN 0.000 0.000 ;
  SYMMETRY X Y R90 ;
  SITE ARM926EJS_site ;
  PIN CLK
    DIRECTION INPUT ;
    USE SIGNAL ;
    PORT
      LAYER METAL2 ;
      RECT 888.620 1122.180 888.820 1123.000 ;
    END
  END CLK
  . . .
  OBS
    LAYER METAL1 ;
      RECT 0.000 0.000 1822.000 1123.000 ;
      RECT 0.000 0.000 1822.000 1123.000 ;
      RECT 0.000 0.000 1822.000 1123.000 ;
    LAYER METAL2 ;
      RECT 699.740 0.000 701.190 1123.000 ;
      RECT 699.530 0.000 701.400 1123.000 ;
      RECT 700.130 0.000 700.800 1123.000 ;

    . . .
  END
END ARM926EJS
. . .
END LIBRARY
```

Figure 3-4   Another LEF file.

Following this are the pin specifications, only one pin specification is
listed in this example, that for pin CLK. It is specified to be an input signal

with the port being defined on METAL2 having the specified rectangle co-ordinates.

In addition to the pin specifications, this macro also has a set of obstructions. These blockages are first defined for METAL1 and then for METAL2.

Quite often, LEF information is split up as a technology file and as a macros file. A technology LEF file contains technology information for a design. This includes placement and routing design rules and process information for the layers. A macros LEF file contains definitions of macros and standard cells for a design.

A technology LEF file can include the following statements.

```
[ VERSION statement ]
[ BUSBITCHARS statement ]
[ DIVIDERCHAR statement ]
[ UNITS statement ]
[ MANUFACTURINGGRID statement ]
[ USEMINSPACING statement ]
[ CLEARANCEMEASURE statement ]
[ PROPERTYDEFINITIONS statement ]
[ LAYER (Nonrouting) statement
  | LAYER (Routing) statement ] ...
[ SPACING statement ]
[ MAXVIASTACK statement ]
[ VIA statement ] ...
[ VIARULE statement ] ...
[ VIARULE GENERATE statement ] ...
[ NONDEFAULTRULE statement ] ...
[ SITE statement ] ...
[ BEGINEXT statement ] ...
[ END LIBRARY ]
```

Figure 3-5 shows an example of a LEF technology file. In the header section, the value of the manufacturing grid has been specified as 0.005u. The next construct defines whether the minimum spacing of blockages (obstructions) are to be honored or not.

```
VERSION 5.4 ;                                   VIARULE VIA1ARRAY GENERATE
NAMESCASESENSITIVE ON ;                            LAYER METAL1 ;
BUSBITCHARS "[ ]" ;                                   DIRECTION HORIZONTAL ;
DIVIDERCHAR "/" ;                                     OVERHANG 0.050 ;
                                                      METALOVERHANG 0.000 ;
UNITS                                              LAYER METAL2 ;
   DATABASE MICRONS 2000  ;                           DIRECTION VERTICAL ;
END UNITS                                             OVERHANG 0.050 ;
MANUFACTURINGGRID 0.005 ;                             METALOVERHANG 0.000 ;
USEMINSPACING OBS OFF ;                            LAYER VIA12 ;
                                                      RECT -0.095 -0.095 0.095 0.095 ;
LAYER POLY1                                           SPACING 0.480 BY 0.480 ;
   TYPE MASTERSLICE ;                           END VIA1ARRAY
END POLY1
                                                VIARULE TURNM1 GENERATE
LAYER METAL1                                        LAYER METAL1 ;
   TYPE ROUTING ;                                    DIRECTION VERTICAL ;
   WIDTH 0.160 ;                                   LAYER METAL1 ;
   SPACING 0.180 ;                                   DIRECTION HORIZONTAL ;
   SPACING 0.18 LENGTHTHRESHOLD  1.0 ;          END TURNM1
   SPACING 0.22 RANGE 0.3 10.0
      USELENGTHTHRESHOLD ;                       SPACING
   SPACING 0.60 RANGE 10.05 100000.0 ;             SAMENET METAL1 METAL1 0.180  ;
   PITCH 0.410 ;                                    SAMENET METAL2 METAL2 0.210  STACK ;
   OFFSET 0.205 ;                                   SAMENET METAL6 METAL6 0.460  ;
   DIRECTION HORIZONTAL ;                           SAMENET VIA12 VIA12 0.220 ;
   AREA 0.122 ;                                     SAMENET VIA23 VIA34 0.0 STACK ;
   MINIMUMCUT 2 WIDTH 1.40 ;                     END SPACING
   THICKNESS 0.26 ;
   ANTENNACUMAREARATIO 5496 ;                    VIA VIA12_H DEFAULT
   ANTENNACUMDIFFAREARATIO PWL                      RESISTANCE 1.0200e+00 ;
      ( ( 0 5496 ) ( 0.159 5496 ) ( 0.16 43062 )    LAYER METAL1 ;
      ( 1 43436 ) ) ;                                  RECT -0.145 -0.105 0.145 0.105 ;
   EDGECAPACITANCE        1.0002e-04 ;             LAYER VIA12 ;
END METAL1                                            RECT -0.095 -0.095 0.095 0.095 ;
                                                   LAYER METAL2 ;
LAYER VIA12                                           RECT -0.145 -0.1 0.145 0.1 ;
   TYPE CUT ;                                    END VIA12_H
   SPACING 0.220 ;
   ANTENNAAREARATIO 50 ;                         VIA VIA23_TOS DEFAULT TOPOFSTACKONLY
   ANTENNADIFFAREARATIO PWL ( ( 0 50 )              RESISTANCE 1.0200e+00 ;
      ( 0.159 50 ) ( 0.16 933 ) ( 1 1110 ) ) ;     LAYER METAL2 ;
END VIA12                                             RECT -0.1 -0.370 0.1 0.370 ;
                                                   LAYER VIA23 ;
LAYER METAL2                                          RECT -0.095 -0.095 0.095 0.095 ;
   TYPE ROUTING ;                                  LAYER METAL3 ;
   WIDTH 0.200 ;                                     RECT -0.145 -0.1 0.145 0.1 ;
   SPACING 0.210 ;                               END VIA23_TOS
   SPACING 0.24 RANGE 0.39 10.0 ;
   SPACING 0.60 RANGE 10.05 100000.0 ;           VIA VIA12_2CUT_W DEFAULT
   PITCH 0.460 ;                                    RESISTANCE 5.1000e-01 ;
   OFFSET 0.230 ;                                   LAYER METAL1 ;
   DIRECTION VERTICAL ;                              RECT -0.625 -0.105 0.145 0.105 ;
   AREA 0.144 ;                                    LAYER VIA12 ;
   MINIMUMCUT 2 WIDTH 1.40 ;                         RECT -0.575 -0.095 -0.385 0.095 ;
   THICKNESS 0.35 ;                                  RECT -0.095 -0.095 0.095 0.095 ;
   ANTENNACUMAREARATIO 5496 ;                     LAYER METAL2 ;
   ANTENNACUMDIFFAREARATIO PWL                       RECT -0.625 -0.1 0.145 0.1 ;
      ( ( 0 5496 ) ( 0.159 5496 ) ( 0.16 43062 )  END VIA12_2CUT_W
      ( 1 43436 ) ) ;                             . . .
END METAL2                                       END LIBRARY
```

Figure 3-5 A technology LEF file.

Following the header is the definition of the various layers. There are five different types of layers that can be specified: masterslice, overlap, cut, implant and routing. The layer descriptions in the LEF file must appear in process order, that is, from bottom to top. For example, POLY (of type masterslice), then METAL1 (of type routing), VIA12 (of type cut), METAL2 (of type routing), VIA23 (of type cut), METAL3 (of type routing), VIA34 (of type cut), and so on.

The first layer in the example is a masterslice layer definition for the POLY1 layer. A masterslice layer is typically a polysilicon layer and is used only if a macro has pins on the polysilicon layer. Following this is the definition of the routing layer METAL1. The default routing width of all routes on this layer is specified to be 0.16u. The minimum spacing between the routes is specified as 0.18u. See Figure 3-6. The length thresh-

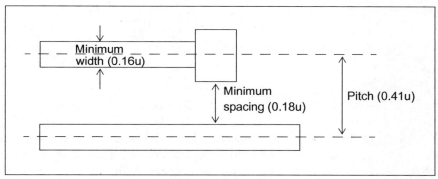

Figure 3-6   Routing layer METAL1.

old of 1u specifies the maximum length a route can run in parallel with an adjacent metal object. The spacing rules apply to any two metal objects, not necessarily with just the routes. The third spacing directive in the example specifies the minimum spacing of 0.22u for objects on the layer with widths in the range of 0.3u to 10u. The USELENGTHTHRESHOLD specifies that this spacing rule is to be applied only if it first meets the LENGTHTHRESHOLD value. The fourth spacing directive specifies the minimum spacing of 0.60u for routes with width of 10.05u to 100000.0u. The pitch specifies the routing pitch for the layer. It defines the routing grid, that is, it defines the distance between routing tracks in the preferred direction. The offset specifies the offset from origin for the routing grid for that layer. This is typically half the pitch, as specified in this case as

0.205u. The direction specifies the preferred routing direction. The area directive specifies the minimum metal area required for polygons on this layer. The thickness directive specifies the thickness of the routes. The edge capacitance is expressed in pf/micron. This is used to estimate capacitance prior to routing.

Next, a cut layer is defined for `VIA12` which is followed by the definition of routing layer `METAL2`. The cut layer is used to specify the vias between layers `METAL1` and `METAL2`. The spacing specification in the definition of `VIA12` layer specifies the minimum spacing between the vias to be 0.22u. See Figure 3-7.

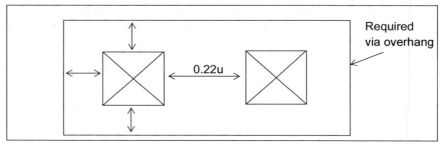

Figure 3-7 Via spacing.

The antenna area ratio of 50 for `VIA12` layer specifies the maximum legal antenna ratio using the area of the metal wire that is not connected to the diffusion diode. The antenna diffusion area ratio specifies the antenna ratio using the area of the metal wire connected to the diffusion diode. The `PWL` specifies a piecewise linear format table by which the ratio is calculated using the linear interpolation of the diffusion area and the ratio input values.

There are two types of vias:

    i.   Fixed
    ii. Generated

Examples of fixed vias are `VIA23_TOS` and `VIA12_H`, and examples of generated vias are `VIA1ARRAY` and `TURNM1`. All vias use three layers: the cut layer and the two layers that it connects. A fixed via is defined using the VIA construct while a generated via is defined using the VI-ARULE...GENERATE statement.

A VIARULE defines vias used at the intersection of special nets such as VSS and VDD. A VIARULE GENERATE defines a recipe or a formula for generating wire arrays. There are two such constructs in this example, for VIA1ARRAY and TURNM1.

The SPACING statement defines spacing between wires of the same net. For example, minimum spacing between METAL1 wires is specified as 0.180u and minimum spacing between METAL6 wires is specified as 0.460u.

The VIA construct defines vias for regular wiring. Three such via definitions appear in the example: VIA12_H, VIA23_TOS and VIA12_2CUT_W.

A macros LEF file contains the following statements.

```
[ VERSION statement ]
[ BUSBITCHARS statement ]
[ DIVIDERCHAR statement ]
[ VIA statement ] ...
[ SITE statement ] ...
[ MACRO statement
  [ PIN statement ] ...
  [ OBS statement ... ] ] ...
[ BEGINEXT statement ] ...
[ END LIBRARY ]
```

Figure 3-8 shows an example of a macros LEF file. Three macros, FILL64, ANTENNA and CLKMX2X12 are defined in this example. The FILL64 macro is defined to be a SPACER cell while the macro ANTENNA is defined to be an ANTENNACELL type cell.

A technology LEF file is typically read in before a macrocell LEF file is read in. The basic rule is that data must be defined before it is used. Thus if you use a site in a LEF file, its definition must precede its use. This rule also dictates the order by which multiple LEF files have to be read in by a LEF reader.

```
VERSION 5.4 ;                              PIN VSS
NAMESCASESENSITIVE ON ;                       DIRECTION INOUT ;
BUSBITCHARS "[ ]" ;                           USE GROUND ;
DIVIDERCHAR "/"  ;                            SHAPE ABUTMENT ;
                                              PORT
SITE TSM13SITE                                   LAYER METAL1 ;
   SYMMETRY Y  ;                                 RECT  0.000 -0.250 0.920 0.250 ;
   CLASS CORE  ;                              END
   SIZE 0.460 BY 3.690 ;                    END VSS
END TSM13SITE                              PIN VDD
                                              DIRECTION INOUT ;
MACRO FILL64                                  USE POWER ;
   CLASS CORE SPACER ;                        SHAPE ABUTMENT ;
   FOREIGN FILL64 0.000 0.000  ;              PORT
   ORIGIN 0.000 0.000 ;                          LAYER METAL1 ;
   SIZE 29.440 BY 3.690 ;                        RECT  0.000 3.440 0.920 3.940 ;
   SYMMETRY X Y ;                             END
   SITE TSM13SITE ;                        END VDD
   PIN VSS                                 END ANTENNA
      DIRECTION INOUT ;
      USE GROUND ;                         MACRO CLKMX2X12
      SHAPE ABUTMENT ;                        CLASS CORE ;
      PORT                                     FOREIGN CLKMX2X12 0.000 0.000  ;
         LAYER METAL1 ;                        ORIGIN 0.000 0.000 ;
         RECT  0.000 -0.250 29.440 0.250 ;     SIZE 8.280 BY 3.690 ;
      END                                      SYMMETRY X Y ;
   END VSS                                     SITE TSM13SITE ;
   PIN VDD                                     PIN Y
      DIRECTION INOUT ;                           DIRECTION OUTPUT ;
      USE POWER ;                                 PORT
      SHAPE ABUTMENT ;                               LAYER METAL1 ;
      PORT                                           RECT  7.685 1.925 7.695 2.585 ;
         LAYER METAL1 ;                              RECT  7.425 1.700 7.685 2.915 ;
         RECT  0.000 3.440 29.440 3.940 ;         END
      END                                         ANTENNADIFFAREA    1.9632 ;
   END VDD                                     END Y
END FILL64                                     PIN S0
                                                  DIRECTION INPUT ;
MACRO ANTENNA                                      PORT
   CLASS CORE ANTENNACELL ;                           LAYER METAL1 ;
   FOREIGN ANTENNA 0.000 0.000  ;                     RECT  0.125 1.585 0.375 2.085 ;
   ORIGIN 0.000 0.000 ;                            END
   SIZE 0.920 BY 3.690 ;                           ANTENNAGATEAREA    0.2457 ;
   SYMMETRY X Y ;                              END S0
   SITE TSM13SITE ;                           PIN VDD
   PIN A                                          DIRECTION INOUT ;
      DIRECTION INPUT ;                           USE POWER ;
      PORT                                        SHAPE ABUTMENT ;
         LAYER METAL1 ;                           PORT
         RECT  0.330 0.875 0.590 2.385 ;             LAYER METAL1 ;
         RECT  0.125 0.875 0.330 1.185 ;             RECT  8.085 3.440 8.280 3.940 ;
      END                                        END VDD
      ANTENNADIFFAREA    1.4270 ;            END CLKMX2X12
   END A                                     . . .
                                             END LIBRARY
```

Figure 3-8   A macros LEF file.

3.3   Complete Syntax

The complete syntax[1] of a LEF file is described in this section. It is described in a non-BNF form. The term `objRegExpr` refers to a regular expression and `pt` represents a point, that is, a coordinate `(x, y)` in the design layout. Three dots "`. . .`" indicate that the previous argument can be repeated. A comma followed by three dots "`, . . .`" indicates that any additional arguments specified must be separated by commas. "`{    }` `. . .`" indicates that the argument must be specified at least once, and can be repeated.

A LEF name cannot contain a new line (\n), a space or a semicolon character. The start of the LEF file syntax is defined by "`LEF file`". The constructs are arranged alphabetically.

```
#Antenna size:
[ ANTENNAINPUTGATEAREA value ; ]
[ ANTENNAINOUTDIFFAREA value ; ]
[ ANTENNAOUTPUTDIFFAREA value ; ]

#Array:
ARRAY arrayName
  { SITE sitePattern ;
  | CANPLACE sitePattern ;
  | CANNOTOCCUPY sitePattern ;
  | TRACKS trackPattern ;
  | FLOORPLAN floorplanName
    { CANPLACE sitePattern | CANNOTOCCUPY sitePattern } ; ...
    END floorplanName } ...
  [ GCELLGRID gcellPattern ; ] ...
  [ DEFAULTCAP tableSize MINPINS numPins WIRECAP cap ; ...
    END DEFAULTCAP]
END arrayName

# Bus bit characters:
BUSBITCHARS "delimiterPair" ;
```

---

1. The syntax is reprinted with permission from "LEF/DEF Language Reference, Product Version 5.6" Copyright Cadence Design Systems, Inc.

```
# Clearance measure:
CLEARANCEMEASURE { MAXXY | EUCLIDEAN } ;

# Divider character:
DIVIDERCHAR "character" ;

# Extensions:
BEGINEXT "tag"
  extension
ENDEXT

# Density statement:
[ DENSITY
  { LAYER layerName ;
    { RECT x1 y1 x2 y2 densityValue ; } ...
  } ...
END ] ...

# Layer (cut):
LAYER LayerName
  TYPE CUT ;
  [ SPACING minSpacing
    [ LAYER 2ndLayerName
      | ADJACENTCUTS { 3 | 4 } WITHIN distance
      | CENTERTOCENTER ]
  ; ] ...
  [ WIDTH minWidth ; ]
  [ ENCLOSURE [ ABOVE | BELOW ] overhang1 overhang2
    [ WIDTH minWidth ] ; ] ...
  [ PREFERENCLOSURE [ ABOVE | BELOW ] overhang1 overhang2
    [ WIDTH minWidth ] ; ] ...
  [ RESISTANCE resistancePerCut ; ]
  [ PROPERTY propName propVal ... ; ] ...
  [ ACCURRENTDENSITY { PEAK | AVERAGE | RMS }
    { value
    | FREQUENCY freq_1 freq_2 ... ;
      [ CUTAREA cutArea_1 cutArea_2 ... ; ]
      TABLEENTRIES
      v_freq_1_cutArea_1 v_freq_1_cutArea_2 ...
      v_freq_2_cutArea_1 v_freq_2_cutArea_2 ...
      ...
    } ; ]
```

```
       [ DCCURRENTDENSITY AVERAGE
         { value
         | CUTAREA cutArea_1 cutArea_2 ... ;
           TABLEENTRIES value_1 value_2 ...
         } ; ]
       [ ANTENNAMODEL { OXIDE1 | OXIDE2 | OXIDE3 | OXIDE4 } ; ] ...
       [ ANTENNAAREARATIO value ; ] ...
       [ ANTENNADIFFAREARATIO { value |
           PWL ( ( d1 r1 ) ( d2 r2 ) ... ) } ; ] ...
       [ ANTENNACUMAREARATIO value ; ] ...
       [ ANTENNACUMDIFFAREARATIO { value |
           PWL ( ( d1 r1 ) ( d2 r2 ) ... ) } ; ] ...
       [ ANTENNAAREAFACTOR value [ DIFFUSEONLY ] ; ] ...
END layerName

# Layer geometries:
{ LAYER layerName
  [ SPACING minSpacing | DESIGNRULEWIDTH value ] ;
    [ WIDTH width ; ]
    { PATH pt ... ;
    | PATH ITERATE pt ... stepPattern ;
    | RECT pt pt ;
    | RECT ITERATE pt pt stepPattern ;
    | POLYGON pt pt pt pt ... ;
    | POLYGON ITERATE pt pt pt pt ... stepPattern ;
    } ...
| VIA pt viaName ;
| VIA ITERATE pt viaName stepPattern ;
} ...

# Layer (implant):
LAYER layerName
  TYPE IMPLANT ;
  [ WIDTH minWidth ; ]
  [ SPACING minSpacing [ LAYER layerName2 ] ; ] ...
  [ PROPERTY propName propVal ; ] ...
END layerName

# Layer (masterslice or overlap):
LAYER layerName
  TYPE { MASTERSLICE | OVERLAP } ;
  [ PROPERTY propName propVal ... ; ] ...
END layerName
```

```
# Layer (routing):
LAYER layerName
  TYPE ROUTING ;
  DIRECTION { HORIZONTAL | VERTICAL | DIAG45 | DIAG135 } ;
  PITCH { distance | xDistance yDistance } ;
  [ DIAGPITCH { distance | diag45Distance diag135Distance } ; ]
  WIDTH defaultWidth ;
  [ OFFSET { distance | xDistance yDistance } ; ]
  [ DIAGWIDTH diagWidth ; ]
  [ DIAGSPACING diagSpacing ; ]
  [ DIAGMINEDGELENGTH diagLength ; ]
  [ AREA minArea ; ]
  [ MINSIZE minWidth minLength [ minWidth2 minLength2 ] ... ; ]
  [ [ SPACING minSpacing
      [ RANGE minWidth maxWidth
        [ USELENGTHTHRESHOLD
          | INFLUENCE value [ RANGE stubMinWidth stubMaxWidth ]
          | RANGE minWidth maxWidth
        ]
        | LENGTHTHRESHOLD maxLength [ RANGE minWidth maxWidth ]
      ]
    ; ] ...
  | [ SPACINGTABLE
      PARALLELRUNLENGTH { length } ...
        { WIDTH width { spacing } ... } ... ;
      { SPACINGTABLE
        INFLUENCE { WIDTH width WITHIN distance
          SPACING spacing } ... ;
    ] ]
  ]
  [ WIREEXTENSION value ; ]
  [ MINIMUMCUT numCuts WIDTH minWidth
    [ FROMABOVE | FROMBELOW ]
    [ LENGTH length WITHIN distance ] ; ] ...
  [ MAXWIDTH width ; ]
  [ MINWIDTH width ; ]
  [ MINSTEP minStepLength
    [ INSIDECORNER | OUTSIDECORNER | STEP ]
    [ LENGTHSUM maxLength ] ; ]
  [ MINENCLOSEDAREA area [ WIDTH width ] ; ] ...
  [ PROTUSIONWIDTH width1 LENGTH length WIDTH width2 ; ]
  [ RESISTANCE RPERSQ value ; ]
  [ CAPACITANCE CPERSQDIST value ; ]
```

```
[ HEIGHT distance ; ]
[ THICKNESS distance ; ]
[ SHRINKAGE distance ; ]
[ CAPMULTIPLIER value ; ]
[ EDGECAPACITANCE value ; ]
[ SLOTWIREWIDTH minWidth ; ]
[ SLOTWIRELENGTH minLength ; ]
[ SLOTWIDTH minWidth ; ]
[ SLOTLENGTH minLength ; ]
[ MAXADJACENTSLOTSPACING spacing ; ]
[ MAXCOAXIALSLOTSPACING spacing ; ]
[ MAXEDGESLOTSPACING spacing ; ]
[ SPLITWIREWIDTH minWidth ; ]
[ MINIMUMDENSITY minDensity ; ]
[ MAXIMUMDENSITY maxDensity ; ]
[ DENSITYCHECKWINDOW windowLength windowWidth ; ]
[ DENSITYCHECKSTEP stepValue ; ]
[ FILLACTIVESPACING spacing ; ]
[ ANTENNAMODEL { OXIDE1 | OXIDE2 | OXIDE3 | OXIDE4 } ; ] ...
[ ANTENNAAREARATIO value ; ]
[ ANTENNADIFFAREARATIO { value |
    PWL ( ( d1 r1 ) ( d2 r2 ) ... ) } ; ]
[ ANTENNACUMAREARATIO value ; ]
[ ANTENNACUMDIFFAREARATIO { value |
    PWL ( ( d1 r1 ) ( d2 r2 ) ... ) } ; ]
[ ANTENNAAREAFACTOR value [ DIFFUSEONLY ] ; ]
[ ANTENNASIDEAREARATIO value ; ]
[ ANTENNADIFFSIDEAREARATIO { value |
    PWL ( ( d1 r1 ) ( d2 r2 ) ... ) } ; ]
[ ANTENNACUMSIDEAREARATIO value ; ]
[ ANTENNACUMDIFFSIDEAREARATIO { value |
    PWL ( ( d1 r1 ) ( d2 r2 ) ... ) } ; ]
[ ANTENNASIDEAREAFACTOR value [ DIFFUSEONLY ] ; ]
[ PROPERTY propName propVal ; ] ...
[ ACURRENTDENSITY { PEAK | AVERAGE | RMS }
  { value
  | FREQUENCY freq_1 freq_2 ... ;
    [ WIDTH width_1 width_2 ... ; ]
    TABLEENTRIES
      v_freq_1_width_1 v_freq_1_width_2 ...
      v_freq_2_width_1 v_freq_2_width_2 ...
      ...
  } ; ]
```

```
            [ DCURRENTDENSITY AVERAGE
              { value
              | WIDTH width_1 width_2 ... ;
                 TABLEENTRIES value_1 value_2 ...
              } ; ]
          END layerName

          #LEF file:
          [ VERSION statement ]
          [ BUSBITCHARS statement ]
          [ DIVIDERCHAR statement ]
          [ UNITS statement ]
          [ MANUFACTURINGGRID statement ]
          [ USEMINSPACING statement ]
          [ CLEARANCEMEASURE statement ]
          [ PROPERTYDEFINITIONS statement ]
          [ LAYER (Nonrouting) statement
            | LAYER (Routing) statement ] ...
          [ SPACING statement ]
          [ MAXVIASTACK statement ]
          [ VIA statement ] ...
          [ VIARULE statement ] ...
          [ VIARULE GENERATE statement ] ...
          [ NONDEFAULTRULE statement ] ...
          [ SITE statement ] ...
          [ MACRO statement
            [ PIN statement ] ...
            [ OBS statement ... ] ] ...
          [ BEGINEXT statement ] ...
          [ END LIBRARY ]

          # Macro:
          MACRO macroName
            [ CLASS
              { COVER [ BUMP ]
              | RING
              | BLOCK [ BLACKBOX | SOFT ]
              | PAD [ INPUT | OUTPUT | INOUT | POWER | SPACER | AREAIO ]
              | CORE [ FEEDTHRU | TIEHIGH | TIELOW
                | SPACER | ANTENNACELL | WELLTAP ]
              | ENDCAP { PRE | POST | TOPLEFT | TOPRIGHT
                | BOTTOMLEFT | BOTTOMRIGHT }
              }
```

```
  ; ]
  [ FOREIGN foreignCellName [ pt [ orient ] ] ; ] ...
  [ ORIGIN pt ; ]
  [ EEQ macroName ; ]
  [ SIZE width BY height ; ]
  [ SYMMETRY { X | Y | R90 } ... ; ]
  [ SITE siteName [ sitePattern ] ; ] ...
  [ PIN statement ] ...
  [ OBS statement ] ...
  [ PROPERTY propName propVal ... ; ] ...
END macroName

# Manufacturing grid:
[ MANUFACTURINGGRID value ; ]

# Maximum via stack:
[ MAXVIASTACK value [ RANGE bottomLayer topLayer ] ; ]

# Names case sensitive:
NAMESCASESENSITIVE { ON | OFF } ;

# Nondefault rule:
[ NONDEFAULTRULE ruleName1
  [ HARDSPACING ; ]
  { LAYER layerName
   WIDTH width ;
    [ DIAGWIDTH diagWidth ; ]
    [ SPACING minSpacing ; ]
    [ WIREXTENSION value ; ]
  END layerName } ...
  [ VIA viaStatement ] ...
  [ USEVIA viaName ; ] ...
  [ USEVIARULE viaRuleName ; ] ...
  [ MINCUTS cutLayerName numCuts ; ] ...
  [ PROPERTY propName propValue ... ; ] ...
END ruleName ]

# No wire extension:
NOWIREXTENSIONATPIN { ON | OFF } ;
```

```
# Macro obstruction statement:
OBS
  { LAYER layerName
    [ SPACING minSpacing | DESIGNRULEWIDTH value ] ;
    [ WIDTH width ; ]
    { PATH pt ... ;
    | PATH ITERATE pt ... stepPattern ;
    | RECT pt pt ;
    | RECT ITERATE pt pt stepPattern ;
    | POLYGON pt pt pt pt ... ;
    | POLYGON ITERATE pt pt pt pt ... stepPattern ;
    } ...
  | VIA pt viaName ;
  | VIA ITERATE pt viaName stepPattern ;
  } ...
END

# Macro pin statement:
PIN pinName
  [ TAPERRULE ruleName ; ]
  [ DIRECTION { INPUT | OUTPUT [ TRISTATE ]
    | INOUT | FEEDTHRU } ; ]
  [ USE { SIGNAL | ANALOG | POWER | GROUND | CLOCK } ; ]
  [ NETEXPR "netExprPropName defaultNetName" ; ]
  [ SUPPLYSENSITIVITY powerPinName ; ]
  [ GROUNDSENSITIVITY groundPinName ; ]
  [ SHAPE { ABUTMENT | RING | FEEDTHRU } ; ]
  [ MUSTJOIN pinName ; ]
  PORT
    [ CLASS { NONE | CORE } ; ]
    layerGeometries ...
  END } ...
  [ PROPERTY propName propVal ... ; ] ...
  [ ANTENNAPARTIALMETALAREA value [ LAYER layerName ] ; ] ...
  [ ANTENNAPARTIALMETALSIDEAREA value
    [ LAYER layerName ] ; ] ...
  [ ANTENNAPARTIALCUTAREA value [ LAYER cutLayerName ] ; ] . . .
  [ ANTENNADIFFAREA value [ LAYER layerName ] ; ] ...
  [ ANTENNAMODEL { OXIDE1 | OXIDE2 | OXIDE3 | OXIDE4 } ; ] . . .
  [ ANTENNAGATEAREA value [ LAYER layerName ] ; ] ...
  [ ANTENNAMAXAREACAR value LAYER layerName ; ] . . .
  [ ANTENNAMAXSIDEAREACAR value LAYER layerName ; ] . . .
```

```
    [ ANTENNAMAXCUTCAR value LAYER layerName ; ] . . .
END pinName

# Property definitions:
[ PROPERTYDEFINITIONS
  [ objectType propName propType [ RANGE min max ]
    [ value | "stringValue" ]
  ; ] ...
END PROPERTYDEFINITIONS ]

# Site:
SITE siteName
  CLASS { PAD | CORE } ;
  [ SYMMETRY { X | Y | R90 } ... ; ]
  [ ROWPATTERN { existingSiteName siteOrient } ... ; ]
  SIZE width BY height ;
END siteName

# Same-net spacing:
[ SPACING
  [ SAMENET
    layerName layerName minSpace [ STACK ] ; ] ...
END SPACING ]

# Units:
[ UNITS
  [ TIME NANOSECONDS convertFactor ; ]
  [ CAPACITANCE PICOFARADS convertFactor ; ]
  [ RESISTANCE OHMS convertFactor ; ]
  [ POWER MILLIWATTS convertFactor ; ]
  [ CURRENT MILLIAMPS convertFactor ; ]
  [ VOLTAGE VOLTS convertFactor ; ]
  [ DATABASE MICRONS convertFactor ; ]
  [ FREQUENCY MEGAHERTZ convertFactor ; ]
END UNITS ]

# Use min spacing:
[ USEMINSPACING OBS { ON | OFF } ; ]

# Version:
VERSION number ;
```

```
# Via:
VIA viaName [ DEFAULT ]
  { VIARULE viaRuleName ;
    CUTSIZE xSize ySize ;
    LAYERS botMetalLayer cutLayer topMetalLayer ;
    CUTSPACING xCutSpacing yCutSpacing ;
    ENCLOSURE xBotEnc yBotEnc xTopEnc yTopEnc ;
    [ ROWCOL numCutRows numCutCols ; ]
    [ ORIGIN xOffset yOffset ; ]
    [ OFFSET xBotOffset yBotOffset xTopOffset yTopOffset ; ]
    [ PATTERN cutPattern ; ]
  | [ RESISTANCE resistValue ; ]
    { LAYER layerName ;
      { RECT pt pt ;
      | POLYGON pt pt pt ... ; } ...
    } ...
  }
  [ PROPERTY propName propVal ; ] ...
END viaName

# Via rule:
VIARULE viaRuleName
  LAYER layerName ;
    DIRECTION { HORIZONTAL | VERTICAL }
    [ WIDTH minWidth TO maxWidth ; ]
  LAYER layerName ;
    DIRECTION { HORIZONTAL | VERTICAL }
    [ WIDTH minWidth TO maxWidth ; ]
  { VIA viaName ; } ...
  [ PROPERTY propName propVal ; ] ...
END viaRuleName

# Via rule generate:
VIARULE viaRuleName GENERATE [ DEFAULT ]
  LAYER routingLayerName ;
    ENCLOSURE overhang1 overhang2
    [ WIDTH minWidth TO maxWidth ; ]
  LAYER routingLayerName ;
    ENCLOSURE overhang1 overhang2
    [ WIDTH minWidth TO maxWidth ; ]
  LAYER cutLayerName ;
    RECT pt pt ;
    SPACING xSpacing BY ySpacing ;
```

    [**RESISTANCE** *resistancePerCut* ;]
END *viaRuleName*

□

4

DEF

T his chapter describes the Design Exchange Format (DEF). It is a
defacto industry standard.

4.1   Basics

DEF allows for the description of the physical design data of a design.
Once a design is physically laid out, a DEF description provides an ASCII
representation of the design. A DEF description also can include logical
design data, such as netlist connectivity. Physical data includes locations
of cells and macros in the design, and locations of routing wires.

A DEF file is generated after a design has been physically placed or
routed by a layout tool. In addition, a DEF file can be read in by a layout

tool in order to perform ECOs (Engineering Change Order) or further layout changes.

A DEF file uses an associated LEF file to provide the entire layout da-
tabase. Thus a DEF file always goes with one or more LEF files. While
the LEF files provide physical information of cells and macros in a library, the DEF file instantiates the cells and macros in its description. Together they thus represent the entire design.

A DEF file along with its LEF files can be used by a parasitic extraction tool to obtain physical information of macrocells being used in a design. Figure 4-1 shows a typical usage of a DEF file.

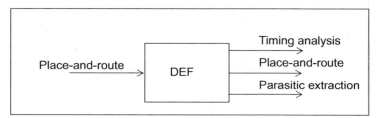

Figure 4-1   DEF is a tool exchange medium.

A DEF file is typically generated by a layout tool as it is specific to a
design.

4.2   Format

Each line in a DEF file is limited to 2048 characters. Each DEF statement ends with a space followed by a semicolon. Statements in a DEF file
can appear in any order, however, data must be defined before it is used.
The format of a DEF file is as follows.

```
[ VERSION statement ]
[ DIVIDERCHAR statement ]
[ BUSBITCHARS statement ]
DESIGN statement
[ TECHNOLOGY statement ]
[ UNITS statement ]
[ HISTORY statement ]
```

```
        [ PROPERTYDEFINITIONS section ]
        [ DIEAREA statement ]
        [ ROWS statement ]
        [ TRACKS statement ]
        [ GCELLGRID statement ]
        [ VIAS statement ]
        [ STYLES statement ]
        [ NONDEFAULTRULES statement ]
        [ REGIONS statement ]
        [ COMPONENTS section ]
        [ PINS section ]
        [ PINPROPERTIES section ]
        [ BLOCKAGES section ]
        [ SLOTS section ]
        [ FILLS section ]
        [ SPECIALNETS section ]
        [ NETS section ]
        [ SCANCHAINS section ]
        [ GROUPS section ]
        [ BEGINEXT section ]
        END DESIGN
```

A DEF file consists of a header followed by a COMPONENTS section and a NETS section that describe the placement and routing of the components and nets in a design. Additional sections such as BLOCKAGES, PINS and SPECIALNETS are often present in a DEF file. The header contains basic information such as version, case sensitivity, hierarchy character, and bit specifier.

Figure 4-2 shows an example[1] of a header definition.

```
        VERSION 5.3 ;
```

specifies the DEF version.

```
        NAMESCASESENSITIVE ON ;
```

specifies that the names are case sensitive.

---

1. The examples shown here are based on DEF versions 5.3 and 5.4, while the grammar and syntax shown are for DEF version 5.6.

```
VERSION 5.3 ;
NAMESCASESENSITIVE ON ;
DIVIDERCHAR "/" ;
BUSBITCHARS  "[ ]" ;
DESIGN DR0012 ;
TECHNOLOGY cl018g_4lm ;
UNITS DISTANCE MICRONS 2000 ;
DIEAREA ( 0 0 ) ( 4680000 4680000 )  ;
ROW row_0 tsm3site 574000 1138480 N DO 2675 BY 1 STEP 1320 10080 ;
ROW row_1 tsm3site 574000 1128400 FS DO 2675 BY 1 STEP 1320 10080
ROW row_2 tsm3site 574000 574000 N DO 2675 BY 1 STEP 1320 10080 ;

. . .
TRACKS X 1780 DO 14181 STEP 330 LAYER METAL1 ;
TRACKS Y 1120 DO 16714 STEP 280 LAYER METAL1 ;
TRACKS X 1780 DO 14181 STEP 330 LAYER METAL2 ;
TRACKS Y 1120 DO 16714 STEP 280 LAYER METAL2 ;

. . .
GCELLGRID X 0 DO 465 STEP 10080  ;
GCELLGRID Y 0 DO 507 STEP 9240  ;
PINS 488 ;
   - nreset
      + NET nreset
      + DIRECTION INPUT
      + USE SIGNAL ;
   - clk_32khz
      + NET clk_32khz
      + DIRECTION INPUT
      + USE SIGNAL ;
   - ntst
      + NET ntst
      + DIRECTION INPUT
      + USE SIGNAL ;
   - pwr_on
      + NET pwr_on
      + DIRECTION INPUT
      + USE SIGNAL ;
   - io[1]
      + NET io[1]
      + DIRECTION INOUT
      + USE SIGNAL ;
   - io[0]
      + NET io[0]
      + DIRECTION INOUT
      + USE SIGNAL ;
   - VDD
      + NET VDD
      + USE POWER
   . . .
END PINS
```

Figure 4-2   A DEF file - first part.

```
DIVIDERCHAR "/" ;
```

specifies the character used in hierarchical DEF names. The backslash character "\" can be used to escape the hierarchy character if it appears in the DEF name itself.

The design name has been specified as DR0012, and the technology library name is c1018g_41m. The UNITS statement:

```
UNITS DISTANCE MICRONS 2000 ;
```

specifies the conversion factor to be used in converting DEF distance units into LEF distance units. The DEF conversion factor should be identical to the LEF conversion factor. The coordinates of the die is specified using the DIEAREA statement:

```
DIEAREA ( 0 0 ) ( 4680000 4680000 )  ;
```

Each ROW statement defines the rows in the design. The ROW statement:

```
ROW row_0 tsm3site 574000 1138480 N
    DO 2675 BY 1 STEP 1320 10080 ;
```

specifies a row with name row_0, followed by the row site type which would be one of the placement sites defined in an associated LEF file. The origin of the first site in the row is specified next followed by the orientation of all sites in the row which is north (N) in this case. Other possible values of orientation are N, S, W, E, FN (F for flip), FS, FW or FE. The DO part specifies the pattern of rows - in this case, make 2675-by-1 rows.

The TRACKS statement defines the routing grid. The statement:

```
TRACKS X 1780 DO 14181 STEP 330 LAYER METAL1 ;
```

defines a set of vertical lines (value X; a value of Y would indicate horizontal lines), with the first line going through the (1780 0) coordinate. The number of tracks is 14181. Spacing between tracks is 330 and the routing layer for the track is METAL1.

A GCELLGRID statement defines the gcell grid. The gcell grid partitions the routable part of the layout into rectangles, called gcells. Each GCELLGRID statement defines a set of vertical (X) lines and a set of horizontal (Y) lines that define the gcell grid. The statements:

```
GCELLGRID X 0 DO 465 STEP 10080 ;
GCELLGRID Y 0 DO 507 STEP 9240 ;
```

create a uniform gcell grid where each gcell is of size 10080 x 9240. The "X 0" defines the origin of the first vertical line, while "Y 0" specifies the origin of the first horizontal line. There are 464 columns in the grid (number specified is one more), and there are 506 rows (number specified is one more).

The PINS statement defines the external pins of the design. It defines the pin name and the net name associated with the pin. For each pin, its attributes such as direction and usage are also specified. In the DEF example shown in Figure 4-2, there are 488 total pins in the design. The first pin is nreset and the net connected to this pin is nreset. It is an input signal pin. Pin io[1] is an example of an inout signal pin. The last pin VDD is an example of a power pin; the net connected to this pin is also called VDD.

Some additional attributes that can be specified for a pin are:

- + **COVER** *pt orient* : Specifies the pin's location which is part of a cover macro. A cover macro cannot be moved by automatic tools or by using interactive commands.

- + **FIXED** *pt orient* : Specifies the pin's location as fixed. It cannot be changed by automatic tools but can be moved interactively.

- + **GROUNDSENSITIVITY** *groundPinName* : Specifies whether the pin is tied low to the specified ground pin name.

- + **LAYER** *layerName ... pt pt* : Specifies the routing layer(s) for the pin.

- + **PLACED** *pt orient* : Specifies the location of the pin. The pin can be moved by automatic tools and interactively.

- + **SPACING** *minSpacing* : Specifies the minimum spacing between this pin and any other routing geometry.

- • + **SUPPLYSENSITIVITY** *powerPinName* : Specifies if the pin is tied high to the specified power pin name.

Figure 4-3 shows the VIAS and COMPONENTS sections of a DEF file.

```
VIAS 183 ;
  - via1Array_DR00121
    + PATTERNNAME via1Array-1.940-0.900-I4
    + RECT VIA12 ( -1820 -780 ) ( -1300 -260 )
    + RECT VIA12 ( -780 -780 ) ( -260 -260 )
    + RECT VIA12 ( 260 -780 ) ( 780 -260 )
    + RECT METAL2 ( -1840 -900 ) ( 1840 900 )
    + RECT METAL1 ( -1940 -800 ) ( 1940 800 ) ;
  - via1Array_DR00122
    + PATTERNNAME via1Array-1.420-0.800-I3
    + RECT VIA12 ( -1300 -780 ) ( -780 -260 )
    + RECT VIA12 ( -260 -780 ) ( 260 -260 )
  - via4Array_DR001277
    + PATTERNNAME via4Array-4.540-0.900-I9
    + RECT METAL4 ( -4440 -900 ) ( 4440 900 ) ;
  . . .
END VIAS

COMPONENTS 15546 ;
  - top_core/qt32s1/qt32c1/BW1_INV39905_7 INVX1
    + PLACED ( 2911720 2549680 ) FN ;
  - top_core/qt32s1/qt32c1/BW1_INV_H24433 INVX1
    + PLACED ( 2196280 2731120 ) FN ;
  - top_core/vid1/vidfifo1/U973_C3_21 AOI22X1
    + PLACED ( 1488760 2499280 ) FS ;
  - top_core/bufc1/U9376_C1_C1 XOR2X1
    + PLACED ( 2957920 2035600 ) FS ;
  - top_core/sdrc1/BW1_INV_H65774 INVX1
    + PLACED ( 3129520 1450960 ) S ;
  - top_core/qt32s1/qt32c1/BL2_BUF288 BUFX1
    + PLACED ( 3211360 2680720 ) FS ;
  - top_core/top_misc1/BL1_ASSIGN_BUF240 BUFX1
    + PLACED ( 1872880 1400560 ) FN ;
  . . .
END COMPONENTS
```

Figure 4-3   A DEF file - continued: VIAS and COMPONENTS.

The VIAS section lists the names and geometries of all vias in the design. In this example, there are 183 vias. Listed are three vias: via1Array_DR00121, via1Array_DR00122, and via4Array_DR001277. The `PATTERNNAME` for each via specifies the encoded value of the cut pattern of the via. It is of the form:

```
viaRuleName-viaSizeX-viaSizeY-cutPattern
```

The `RECT` statements define the via geometry for the specified layer. In the example, the via `via1Array_DR00121` is defined to have five `RECT` statements, first three on layer `VIA12` and the others on `METAL2` and `METAL1`.

The cut pattern format in DEF version 5.6 is different from version 5.3. In 5.6, the cut pattern is of the form:

```
+ PATTERN numRows1_rowDef1_numRows2_rowDef2_ . . . _
numRowN_rowDefN;
```

where the total number of rows must match the number of via rows specified by the `ROWCOL` attribute. Here is an example.

```
- via6Array_DR0021
  + ROWCOL 3 5
  + PATTERN 1_78_2_E0 ;
```

Figure 4-4 shows the via pattern for the above example. There are three

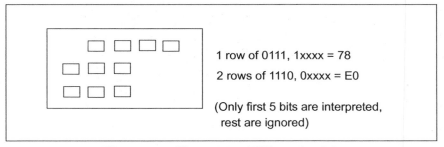

1 row of 0111, 1xxxx = 78

2 rows of 1110, 0xxxx = E0

(Only first 5 bits are interpreted,
  rest are ignored)

Figure 4-4   Via pattern, DEF version 5.6.

rows and five columns. The first row has a pattern of `78` while the next two rows have the pattern of `E0`.

In DEF version 5.3, there are two forms of pattern cut encoding.

i. Bit-mapped

ii. Compressed

In bit-mapped encoding, the cut pattern is represented as hexadecimal digits. This is shown in an example in Figure 4-5. The via cut pattern for

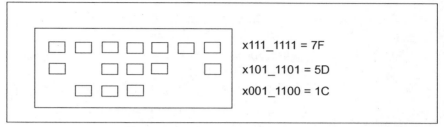

Figure 4-5 Bit-mapped encoding.

this example is:

```
viaRuleName-viaSizeX-viaSizeY-7F-5D-1C
```

In the compressed notation form, both hexadecimal and double-decimal (base 20) digits are used to represent the cut patterns. In double-decimal, the character set consists of G to Z with values sequentially ranging from 0 to 19 - G has value 0 and Z has value 19. This notation also supports the writing of repetitive patterns. Such a cut pattern is of the form:

```
-RpcutPatternRh
```

R_p is the number of times in double-decimal that the cut pattern repeats itself. R_h is the number of times the preceding character repeats. Here is an example.

```
- via8Array_DR0022
  + PATTERNNAME via8Array-1.0-1.5-2K53FJ
```

The cut pattern is `53FFFF` (the last character repeats 3 more times) and the pattern repeats `2K` (in decimal 44) times.

Some additional attributes that can be specified for a via are:

- + **CUTSIZE** *xSize ySize* : Specifies the width and height of the cut layer rectangles.
- + **CUTSPACING** *xCutSpacing yCutSpacing* : Specifies the spacing between the cuts.
- + **LAYERS** *botMetalLayer cutLayer TopMetalLayer* : Specifies the layers that comprise the cut.
- + **POLYGON** *layerName pt pt pt* : Specifies the via geometry on the specified layer.
- + **ROWCOL** *numCutRows numCutCols* : Specifies the number of rows and columns in the cut array.

The COMPONENTS section of the DEF file defines all the components in the design. For each component, the instance name, the component type, its placement status and coordinates, if any, followed by the orientation of the component are specified. In the example of Figure 4-3, there are a total of 15546 components in the design. The first component instance name is top_core/qt32s1/qt32c1/BW1_INV39905_7 and the component type is INVX1. The instance is placed at location (2911720 2549680) and its orientation is FN (flip north). Other choices of orientation values are N, E, S, W, FE (flip east), FS (flip south), and FW (flip west).

Additional attributes that can be specified for a component are:

- + **COVER** *pt orient* : Specifies the component to be part of a cover macro. A cover macro cannot be moved by automatic or interactive commands.
- + **EEQMASTER** *macroName* : Specifies the component is electrically equivalent to the macro.
- + **FIXED** *pt orient* : Specifies that the component is fixed. It cannot be moved by automatic tools but can be moved interactively.
- + **HALO** *left bottom right top* : Specifies a placement blockage the specified distance away from the edges.
- + **PROPERTY** *propName propVal* : Specifies the value of a property that has been defined earlier in the PROPERTYDEFINITIONS statement.

- + **REGION** *regionName* : Specifies the region in which the component is placed.

- + **SOURCE** { NETLIST | DIST | USER | TIMING } : Specifies how the component was created.

- + **UNPLACED**

- + **WEIGHT** *weight* : Specifies the weightage used in placing the component at the specified location.

Figure 4-6 shows the SPECIALNETS section of the DEF file. The

```
SPECIALNETS 40 ;
 - top_core/xmem_done
   ( top_core/bufc1/U3429_C2_2 B1 )
   ( top_core/bufc1/U4098_C2_1 A1N )
   ( top_core/bufc1/U6658_C2_2 B1 )
   ( top_core/sdrc1/xmem_done_reg Q )
     + SOURCE NETLIST
     + USE SIGNAL
     + ROUTED METAL1 460 ( 3362600 1628480 ) ( 3364710 * )
       NEW METAL1 460 ( 3153710 1640800 ) ( 3154500 * )
       NEW METAL2 560 ( 3153940 1640520 ) ( * 1644440 )
       NEW METAL2 560 ( 3337420 1628200 ) ( * 1644440 )
       NEW METAL2 560 ( 3364480 1627870 ) ( * 1629320 )
 - top_core/qt32s1/qt32c1/new_bit_cnt\[3\]
   ( top_core/qt32s1/qt32c1/add_2364_plus_plus_U1_3 S )
   ( top_core/qt32s1/qt32c1/U10857_C3_1 A0 )
   ( top_core/qt32s1/qt32c1/BW1_INV_H23777 A )
     + SOURCE NETLIST
     + USE SIGNAL
     + ROUTED METAL1 460 ( 2232350 2823520 ) ( 2237760 * )
       NEW METAL1 460 ( 2231260 2767570 ) ( * 2769710 )
       NEW METAL1 460 ( 2225750 2769480 ) ( 2231490 * )
       NEW METAL1 460 ( 2240500 2814790 ) ( * 2819550 )
       NEW METAL1 920 ( 2240730 2815350 ) ( * 2818030 )
       NEW METAL1 460 ( 2237300 2823800 ) ( 2239410 * )
       NEW METAL1 460 ( 2238950 2819320 ) ( 2240730 * )
       NEW METAL1 790 ( 2240665 2814790 ) ( * 2818030 )
   . . .
 END SPECIALNETS
```

Figure 4-6   A DEF file - continued: SPECIALNETS.

SPECIALNETS section defines connectivity of nets that contain special pins, such as power nets and shield nets. Typically such nets need special handling and are not routed by the standard router. In our example, there are a total of 40 special nets, two of which are shown in the figure. The connectivity of the net is shown after the net name - the sources and the sinks, with instance name and pin name. The SOURCE statement specifies that the net was created from a netlist. Other choices for the SOURCE are DIST (added new cells), TIMING (net created due to a logical change), and USER (net is user-defined). The USE statement specifies how the net is used. Values for this can be ANALOG, CLOCK, GROUND, POWER, RESET, SCAN, SIGNAL or TIEOFF. This is followed by the net route segment descriptions (ROUTED statement). For each route segment, its layer is specified with the width of the segment, followed by the coordinates of the rectangle. The * indicates "keep previous value". So the following segment description:

```
NEW METAL1 460 ( 3153710 1640800 )  ( 3154500 * )
```

specifies the route segment to be on METAL1 with segment thickness to be 460, and that it is a horizontal route (Y does not change).

Figure 4-7 shows the last part of the DEF file. This figure shows the NETS section. This section defines the connectivity of nets that contain regular pins. Together with the SPECIALNETS section, they describe the entire connectivity of the design. Each of the nets is described in the same form as the nets in the SPECIALNETS section. In our example, there are 35639 regular nets in the design. The last net listed is:

```
top_core/qt32s1/qt32c1/r2657_U551_C1__n_3
```

and is connected to the following pins:

```
( top_core/qt32s1/qt32c1/r2657_U357_C1_1 Y )
( top_core/qt32s1/qt32c1/BW2_INV27360 A )
( top_core/qt32s1/qt32c1/r2657_U511_C3_2 B0 )
```

The net was created from a netlist (SOURCE statement) and is a signal net and the route is composed of seven segments, five on METAL1 and two on METAL2.

```
NETS 35639 ;
  - sdr_data[0]
    ( PIN sdr_data[0] )
    ( io_pads/sdrd0_p_U6 PAD )
      + SOURCE NETLIST
      + USE SIGNAL ;
  - sdr_data[1]
    ( PIN sdr_data[1] )
    ( io_pads/sdrd1_p_U6 PAD )
      + SOURCE NETLIST
      + USE SIGNAL ;
  - sdr_data[2]
    ( PIN sdr_data[2] )
    ( io_pads/sdrd2_p_U6 PAD )
      + SOURCE NETLIST
      + USE SIGNAL ;
  - top_core/qt32s1/qt32c1/r2657_U551_C1__n_3
    ( top_core/qt32s1/qt32c1/r2657_U357_C1_1 Y )
    ( top_core/qt32s1/qt32c1/BW2_INV27360 A )
    ( top_core/qt32s1/qt32c1/r2657_U511_C3_2 B0 )
      + SOURCE NETLIST
      + USE SIGNAL
      + ROUTED METAL1 ( 2986300 3419920 ) ( 2986630 * )
        NEW METAL1 ( 2986630 3419920 ) ( * 3420760 )
        NEW METAL1 ( 2986630 3420760 ) ( 2987620 * )
        NEW METAL1 ( 2986630 3420540 ) ( * 3420760 )
        NEW METAL1 ( 2987090 3420540 ) ( * 3420760 )
        NEW METAL2 ( 2973100 3427200 ) ( * 3449060 )
        NEW METAL2 ( 2971450 3449060 ) ( * 3451660 )
  . . .
END NETS

END DESIGN
```

Figure 4-7   A DEF file - last part.

Here are some additional attributes that a net can have.

- + **ESTCAP** *wireCapacitance* : Specifies the estimated wire capacitance for the net.

- + **FREQUENCY** *frequency* : Specifies the maximum frequency of the net.

- + **MUSTJOIN** (*compName pinName*) : Specifies that the net is a mustjoin, in such a case, the net name is automatically generated by the system. A mustjoin net is used to specify any prewiring.

- + **NONDEFAULTRULE** *ruleName* : Specifies the nondefault rule defined in the LEF file to be used for the net.

- + **ORIGINAL** *netName* : Specifies the original name of the net that was used to create multiple nets including the one being defined.

- + **PATTERN** { BALANCED · | STEINER | TRUNK | WIREDLOG-IC } : Specifies the routing style to be used for the net.

- + **PROPERTY** *propName propVal* : Specifies the value for a property that has been defined earlier using the PROPERTYDEFINI-TIONS statement.

- + **SHIELDNET** *shieldNetName* : Specifies the name of the shield net that shields the net being defined.

- + **WEIGHT** *weight* : Specifies the weight of the net.

- + **XTALK** *class* : Specifies the crosstalk class number for the net.

4.3   Complete Syntax

The syntax[1] of a DEF file is described in this section. It is described in a non-BNF form. The term objRegExpr refers to a regular expression and pt represents a point, that is, a coordinate (x, y) in the design. Three dots " . . . " indicate that the previous argument can be repeated. A comma followed by three dots ", . . .", indicates that any more arguments specified must be separated by commas.

A DEF name cannot contain a new line (\n), a space or a semicolon character. The start of the DEF file syntax is defined by "DEF file".

```
#Array:
ARRAY arrayName ;

#Blockages:
[ BLOCKAGES numBlockages ;
  [ - LAYER layerName
    [ + COMPONENT compName | + SLOTS | + FILLS | + PUSHDOWN ]
    [ + SPACING minSpacing | + DESIGNRULEWIDTH effectiveWidth ]
```

---

1. The syntax is reprinted with permission from "LEF/DEF Language Reference, Product Version 5.6" Copyright Cadence Design Systems, Inc.

```
        { RECT pt pt | POLYGON pt pt pt ... } ...
   ; ] ...
   [ - PLACEMENT
     [ + COMPONENT compName | + PUSHDOWN ]
        { RECT pt pt } ...
   ; ] ...
 END BLOCKAGES ]
```

```
# Bus bit characters:
BUSBITCHARS "delimiterPair" ;
```

```
# Cannot occupy:
CANNOTOCCUPY
  sitename origX origY orient
    DO numX BY numY STEP spaceX spaceY ;
```

```
# Can place:
CANPLACE
  sitename origX origY orient
    DO numX BY numY STEP spaceX spaceY
```

```
# Components:
COMPONENTS numComps ;
  [ - compName modelName
    [ + EEQMASTER macroName ]
    [ + SOURCE { NETLIST | DIST | USER | TIMING } ]
    [ + { FIXED pt orient | COVER pt orient | PLACED pt orient
      | UNPLACED } ]
    [ + HALO left bottom right top ]
    [ + WEIGHT weight ]
    [ + REGION regionName ]
    [ + PROPERTY { propName propVal } ... ] ...
  ; ] ...
END COMPONENTS
```

```
# DEF file:
[ VERSION statement ]
[ DIVIDERCHAR statement ]
[ BUSBITCHARS statement ]
DESIGN statement
[ TECHNOLOGY statement ]
[ UNITS statement ]
[ HISTORY statement ] ...
```

```
                [ PROPERTYDEFINITIONS section ]
                [ DIEAREA statement ]
                [ ROWS statement ] ...
                [ TRACKS statement ] ...
                [ GCELLGRID statement ] ...
                [ VIAS statement ]
                [ STYLES statement ]
                [ NONDEFAULTRULES statement ]
                [ REGIONS statement ]
                [ COMPONENTS section ]
                [ PINS section ]
                [ PINPROPERTIES section ]
                [ BLOCKAGES section ]
                [ SLOTS section ]
                [ FILLS section ]
                [ SPECIALNETS section ]
                [ NETS section ]
                [ SCANCHAINS section ]
                [ GROUPS section ]
                [ BEGINEXT section ] ...
                END DESIGN statement

                # Design:
                DESIGN designName ;

                # Die area:
                [ DIEAREA pt pt [ pt ] ... ; ]

                # Divider character:
                DIVIDERCHAR "character" ;

                # Extensions:
                [ BEGINEXT "tag"
                  extensionText
                ENDEXT ]

                # Fills:
                [ FILLS numFills ;
                  [ - LAYER layerName
                    { RECT pt pt | POLYGON pt pt pt ... } ...
                  ; ] ...
                END FILLS ]
```

```
# Floorplan:
FLOORPLAN { DEFAULT | floorPlanName } ;

# GCell grid:
[ GCELLGRID
  { X start DO numColumns+1 STEP space } ...
  { Y start DO numRows+1 STEP space ;} ... ]

# Groups:
[ GROUPS numGroups ;
  [ - groupName compNameRegExpr ...
    [ + REGION regionName ]
    [ + PROPERTY { propName propVal } ... ] ...
  ; ] ...
END GROUPS ]

# History:
[ HISTORY anyText ; ] ...

# Names case sensitive:
NAMESCASESENSITIVE { OFF | ON } ;

# Nets:
NETS numNets ;
  [ - { netName
    [ ( { compName pinName | PIN pinName }
      [ + SYNTHESIZED ] ) ] ...
      | MUSTJOIN ( compName pinName ) }
    [ + SHIELDNET shieldNetName ] ...
    [ + VPIN vpinName [ LAYER layerName ] pt pt
      [ PLACED pt orient | FIXED pt orient |
        COVER pt orient ] ] ...
    [ + SUBNET subnetName
      [ ( { compName pinName | PIN pinName |
        VPIN vpinName } ) ] ...
      [ NONDEFAULTRULE ruleName ]
      [ regularWiring statement ] ... ] ...
    [ + XTALK class ]
    [ + NONDEFAULTRULE ruleName ]
    [ regularWiring statement ] ...
    [ + SOURCE { DIST | NETLIST | TEST | TIMING | USER } ]
    [ + FIXEDBUMP ]
    [ + FREQUENCY frequency ]
```

```
      [ + ORIGINAL netName ]
      [ + USE { ANALOG | CLOCK | GROUND | POWER | RESET
        | SCAN | SIGNAL | TIEOFF } ]
      [ + PATTERN { BALANCED | STEINER | TRUNK | WIREDLOGIC } ]
      [ + ESTCAP wireCapacitance ]
      [ + WEIGHT weight ]
      [ + PROPERTY { propName propVal } ... ] ...
    ; ] ...
END NETS

# Regular wiring statement:
{ + COVER | + FIXED | + ROUTED | + NOSHIELD }
  layerName [ TAPER | TAPERRULE ruleName ] [ STYLE styleNum ]
   ( x y [ extValue ] )
   { ( x y [ extValue ] ) | viaName [ orient ] } ...
  | NEW layerName [ TAPER | TAPERRULE rulename ]
   [ STYLE styleNum ]
   ( x y [ extValue ] )
   { ( x y [ extValue ] ) | viaName [ orient ] } ...
  ] ...

# Nondefault rules:
NONDEFAULTRULES numRules ;
  { - ruleName
    [ + HARDSPACING ]
    { + LAYER layerName
      WIDTH minWidth
      [ DIAGWIDTH diagWidth ]
      [ SPACING minSpacing ]
      [ WIREEXT wireExt ]
    } ...
    [ + VIA viaName ]
    [ + VIARULE viaRuleName ] ...
    [ + MINCUTS cutLayerName numCuts ] ...
    [ + PROPERTY { propName propVal } ... ] ...
  ; } ...
END NONDEFAULTRULES

# Pins:
[ PINS numPins ;
  [ [ - pinName + NET netName ]
    [ + SPECIAL ]
    [ + DIRECTION { INPUT | OUTPUT | INOUT | FEEDTHRU } ]
```

```
          [ + NETEXPR "netExprPropName defaultNetName" ]
          [ + SUPPLYSENSITIVITY powerPinName ]
          [ + GROUNDSENSITIVITY groundPinName ]
          [ + USE { SIGNAL | POWER | GROUND | CLOCK | TIEOFF
            | ANALOG | SCAN | RESET } ]
          [ + ANTENNAPINPARTIALMETALAREA value
            [ LAYER layerName ] ] ...
          [ + ANTENNAPINPARTIALMETALSIDEAREA value
            [ LAYER layerName ] ] ...
          [ + ANTENNAPINPARTIALCUTAREA value [ LAYER layerName ] ] ...
          [ + ANTENNAPINDIFFAREA value [ LAYER layerName ] ] ...
          [ + ANTENNAMODEL { OXIDE1 | OXIDE2 | OXIDE3 | OXIDE4 } ] ...
          [ + ANTENNAPINGATEAREA value [ LAYER layerName ] ] ...
          [ + ANTENNAPINMAXAREACAR value LAYER layerName ]
          [ + ANTENNAPINMAXSIDEAREACAR value LAYER layerName ]
          [ + ANTENNAPINMAXCUTCAR value LAYER layerName ]
          [ + LAYER layerName
            [ SPACING minSpacing | DESIGNRULEWIDTH effectiveWidth ]
            pt pt ]
          | + POLYGON layerName
            [ SPACING minSpacing | DESIGNRULEWIDTH effectiveWidth ]
            pt pt pt ... ] ...
          [ + COVER pt orient | FIXED pt orient
            | PLACED pt orient ]
        ; ] ...
    END PINS ]

# Pin properties:
[ PINPROPERTIES num ;
   [ - { compName pinName | PIN pinName }
     [ + PROPERTY propName propVal ... ] ...
   ; ] ...
END PINPROPERTIES ]

# Property definitions:
[ PROPERTYDEFINITIONS
   [ objectType propName propType [ RANGE min max ]
     [ value | stringValue ]
   ; ]...
END PROPERTYDEFINITIONS ]
```

```
# Regions:
[ REGIONS numRegions ;
  [ - regionName { pt pt } ...
    [ + TYPE { FENCE | GUIDE } ]
    [ + PROPERTY { propName propVal } ... ] ...
  ; ] ...
END REGIONS ]

# Routing points:
( x y [ extValue ] )
  { ( x y [ extValue ] )
  | viaName
      [ DO numX BY numY STEP stepX stepY ]
  } ...
DO numX BY numY STEP stepX stepY

# Rows:
[ ROW rowName siteName origX origY siteOrient
  [ DO numX BY numY [ STEP stepX stepY ] ]
  [ + PROPERTY { propName propVal } ... ] ... ; ] ...

# Scan chains:
[ SCANCHAINS numScanChains ;
  [ - chainName
    [ + PARTITION partitionName [ MAXBITS maxBits ] ]
    [ + COMMONSCANPINS [ ( IN pin ) ] [ ( OUT pin ) ] ]
      + START { fixedInComp | PIN } [ outPin ]
    [ + FLOATING
      { floatingComp [ ( IN pin ) ] [ ( OUT pin ) ]
        [ ( BITS numBits ) ] } ... ]
    [ + ORDERED
      { fixedComp [ ( IN pin ) ] [ ( OUT pin ) ]
        [ ( BITS numBits ) ] } ...
    ] ...
      + STOP { fixedOutComp | PIN } [ inPin ] ]
  ; ] ...
END SCANCHAINS ]

# Slots:
[ SLOTS numSlots ;
  [ - LAYER layerName
    { RECT pt pt | POLYGON pt pt pt ... } ...
```

```
    ; ] ...
END SLOTS ]

# Special nets:
[ SPECIALNETS numNets ;
  [ - netName [ ( compNameRegExpr pinName
      [ + SYNTHESIZED ] ) ] ...
    [ + VOLTAGE volts ]
    [ specialWiring ] ...
    [ + SOURCE { DIST | NETLIST | TIMING | USER } ]
    [ + ORIGINAL netName ]
    [ + USE { ANALOG | CLOCK | GROUND | POWER | RESET
      | SCAN | SIGNAL | TIEOFF } ]
    [ + PATTERN { BALANCED | STEINER | TRUNK | WIREDLOGIC } ]
    [ + ESTCAP wireCapacitance ]
    [ + WEIGHT weight ]
    [ + PROPERTY { propName propVal } ... ] ...
  ; ] ...
END SPECIALNETS ]

# Special wiring statement:
[ + POLYGON layerName pt pt pt ...
  | + RECT layerName pt pt
  | { + COVER | + FIXED | + ROUTED | + SHIELD shieldNetName }
    layerName routeWidth
      [ + SHAPE { RING | PADRING | BLOCKRING | STRIPE | FOLLOWPIN
        | IOWIRE | COREWIRE | BLOCKWIRE | BLOCKAGEWIRE
        | FILLWIRE | DRCFILL } ]
      [ + STYLE styleNum ]
      routingPoints
    [ NEW layerName routeWidth
      [ + SHAPE { RING | PADRING | BLOCKRING | STRIPE | FOLLOWPIN
        | IOWIRE | COREWIRE | BLOCKWIRE | BLOCKAGEWIRE
        | FILLWIRE | DRCFILL } ]
      [ + STYLE styleName ]
      routingPoints
    ] ...
  ] ...

# Styles:
[ STYLES numStyles ;
  { - STYLE styleNum pt pt pt ... ; } ...
END STYLES ]
```

```
# Technology:
[ TECHNOLOGY technologyName ; ]

# Tracks:
[ TRACKS
  [ { X | Y } start DO numtracks STEP space
    [ LAYER layerName ] ...
  ; ] ... ]

# Units:
[ UNITS DISTANCE MICRONS dbuPerMicron ; ]

# Version:
[ VERSION versionNumber ; ]

# Vias:
[ VIAS numVias ;
  [ - viaName
    [ + VIARULE viaRuleName
      + CUTSIZE xSize ySize
      + LAYERS botMetalLayer cutLayer topMetalLayer
      + CUTSPACING xCutSpacing yCutSpacing
      + ENCLOSURE xBitEnc yBotEnc xTopEnc yTopEnc
      [ + ROWCOL numCutRows numCutCols ]
      [ + ORIGIN xOffset yOffset ]
      [ + OFFSET xBotOffset yBotOffset xTopOffset yTopOffset ]
      [ + PATTERN cutPattern ] ]
    | [ + RECT layerName pt pt
      | + POLYGON layerName pt pt pt ] ... ]
  ; ] ...
END VIAS ]
```

PDEF

T his chapter describes the Physical Design Exchange Format
(PDEF). It is part of the IEEE Std 1481-1999.

5.1   Basics

PDEF provides a standard mechanism for passing physical design information between high-level design tools and physical design tools. The
key ability is that it provides support for descriptions of physical partitions of design that is independent of the logical hierarchy. This physical
design information is presented in an ASCII form in a PDEF file.

A PDEF file for a design supports the description of mapping of physical groupings to the logical hierachy, including description of physical
constraints. It may contain, in addition, description of wireload models

and global routing information. A PDEF file, however, does not contain logical connectivity, delay information, detailed routing or parasitic information.

PDEF is used to describe the physical view (physical layout) of a design. Typically, there are fewer physical partitions than logical levels of hierarchy. A PDEF file captures the mapping of the physical partitions to the logical hierarchies. It is also possible for a physical partition to cross logical hierarchies. Physical partitions can also be nested.

A PDEF file is used to pass physical, grouping and net information between frontend and backend tools. It also defines components and nets used in the design with their physical attributes. Power domains can be explicitly described and there is provision for describing special pre-routed blocks. Mapping of groupings and partitions to logical hierarchy can be explicitly described. Design-specific definitions, such as wireload models and interconnect layers, can be described. A PDEF file also supports power and delay calculation - a tool can read in this file and perform delay and power calculation.

A PDEF file is typically generated by a high-level design tool such as a floorplanner, placement tool, or a partitioning tool. A PDEF file can be read by a place and route tool to continue routing, or by a delay calculator to estimate interconnect delays, or by a floorplanner for making changes. Figure 5-1 shows a typical usage of a PDEF file.

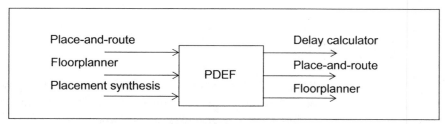

Figure 5-1   PDEF is a tool exchange medium.

A physical partition is described by a cluster in PDEF. A physical partition, that is, a cluster, can encompass one or more logical hierarchies. Thus the physical information of an entire design is represented by a top cluster with one or more subclusters, with each subcluster having additional subclusters hierarchically. See Figure 5-2. The logical hierarchy is shown on the left and the physical layout is shown on the right. The top-level cluster includes three clusters, C1, C2, C3, with cluster C3 having an additional subcluster C4. The logical hierarchy shows module M1 containing logic modules M2 and M3, M2 containing M4 and M7, and so on. Logic subhierarchy M2 is assigned to cluster C1, logic subhierarchy M5 is assigned to cluster C2 while logic modules M3 and M6 are assigned to cluster C3, with M6 in its own subcluster C4. The coordinates for the lower-left corner and the upper-right corner of the clusters are also shown in the figure.

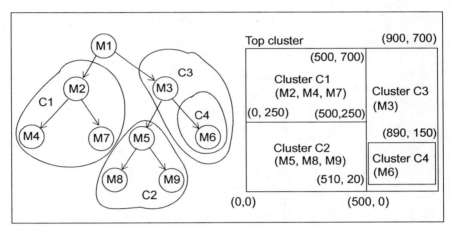

Figure 5-2   A cluster can encompass one or more logical hierarchies.

It is possible for a cluster to cross hierarchical boundaries as well. See Figure 5-3. Cluster C2 contains modules M5, M7 and M8 which are from different hierarchical blocks. Similarly, modules M6 and M9, which are in different hierarchical blocks, are in cluster C4.

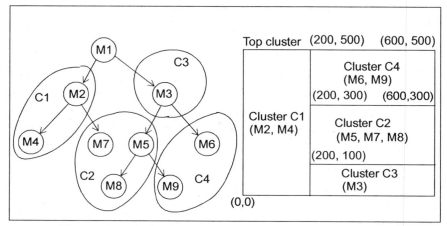

Figure 5-3 A cluster may cross logical hierarchies as well.

5.2   Format

The format of a PDEF file is as follows.

```
(CLUSTERFILE
header_definition
[ wire_load_definition ]
[ layer_definition ]
[ via_definition ]
[ gate_definition ]
{ cluster } )
```

The *header definition* contains basic information such as PDEF version number, design name, pin and bus delimiters, and hierarchy delimiter. The header contains information about the design, the PDEF file and the design flow. The header is followed by descriptions of *design-specific components* such as wireload models, abstracts, and routing layers. The wireload definition specifies capacitance and resistance tables for wires that are based on number of pins in the net. The *layer definition* provides the description of the various physical layers. The *via definition* section is used to specify all vias used in the physical design. The *gate definition* specifies the definitions of any abstract gates used in the design. The *cluster* section contains the real meat of the PDEF file, which are the descriptions of the physical partitions and groupings in the design. Each cluster contains description of cells that belong to the cluster, description of nets that belong to the cluster and information on any global routes, and description of additional subclusters, if any.

Figure 5-4 shows an example of a header definition. The PDEF ver-

```
(CLUSTERFILE
   (PDEFVERSION "IEEE 1481-1999")
   (DESIGN DR2119)
   (DATE "Wed Jan  5 08:44:05 2005")
   (VENDOR "Star Galaxy Automation, Inc")
   (PROGRAM "PlannerRTL")
   (DIVIDER / )
   (PIN_DELIMITER / )
   (BUS_DELIMITER  [ ])
   (NETLIST_TYPE VERILOG)
   (DESIGN_FLOW cell_locations)
   (DESIGN_FLOW gates)
   (NAMEPREFIX 1 i_takpipe)
   (NAMEPREFIX 2 i_ram2048x32/wrapper)
   (RESISTANCE_UNIT 0.1)
   (CAPACITANCE_UNIT 1.0)
   (DISTANCE_UNIT 0.1)
```

Figure 5-4   A header definition of a design.

sion is specified as IEEE1481-1999. The design name is DR2119. The hierarchy divider character is the '/' character. Any pin specification is delimited by the '/' character and any bus bits specified are enclosed within " [] ".

The NETLIST_TYPE is specified as VERILOG; this indicates that Verilog HDL naming conventions are used in the PDEF file. Other possible values for this construct are:

- VHDL87: Uses VHDL87 naming conventions.
- VHDL93: Uses VHDL93 naming conventions.
- EDIF: Uses EDIF naming conventions.

The DESIGN_FLOW construct specifies the kind of information that is present in the PDEF file. The possible values for this construct are:

- cell_locations: Locations are defined for some or all cells, using LOC or X_BOUNDS and Y_BOUNDS.
- cluster_locations: Locations are specified for clusters, using RECT or X_BOUNDS and Y_BOUNDS.
- pin_locations: Locations are specified for some or all pins, using RECT or LOC.
- datapath_constraints: Placement constraints, such as ROW, COL, H_POS, V_POS for pin and net placement and GAP for cluster placement, are present in the PDEF file.
- datapath_placement: CONTENT_ORDER is used to determine cluster and cell placement.
- gates: Gate definitions are present in the PDEF file.
- logical_clusters: Clusters correspond to a logical hierarchy.
- obstruction: OBSTRUCTION is present in PDEF file.
- power: POWER domains are defined.
- routes: A net or a pnet has route information specified.
- complete_routes: All routes described have complete connectivity.

There can be multiple DESIGN_FLOW constructs in a file and all information is cumulative.

The NAMEPREFIX construct can be used to provide an alias for a long name. For example, index 2 is an alias for i_ram2048x32/wrapper. So where ever a name is expected, the index of NAMEPREFIX can also be provided. This helps in reducing the amount of text, and thus the size of the PDEF file.

The design-specific components for an example design are shown in Figure 5-5. These consist of wireload definitions, gate definitions, via definitions and routing layer definitions.

A wireload definition provides a pincount-based model for estimating wire resistance and wire capacitance. In this example, two wireload mod-

```
(WIRE_LOAD_DEF
  (WIRE_LOAD light_wl
    (PINCOUNT 2 (RES 1.0) (CAP 0.2))
    (PINCOUNT 3 (RES 1.5) (CAP 0.4))
    (PINCOUNT 4 (RES 2.0) (CAP 0.6))
  )
  (WIRELOAD heavy_wl
    (PINCOUNT 2 (RES 1.1) (CAP 0.3))
    (PINCOUNT 3 (RES 1.6) (CAP 0.5))
    (PINCOUNT 5 (RES 2.7) (CAP 0.91))
  )
)
(LAYER_DEF
  (LAYER METAL1 1
    (RES 0.011) (CAP 0.003) (ORIENT "0")
    (ROUTING_PERCENTAGE 30) (GAP 0.21)
    (WIDTH 0.2300)
  )
  (LAYER METAL1 2 (WIDTH 0.4600) )
  (LAYER METAL2 19 (WIDTH 0.2800) )
  (LAYER METAL2 20 (WIDTH 0.4950) )
  (LAYER METAL3 31 (WIDTH 0.2800) )
  (LAYER METAL3 32 (WIDTH 0.4200) )
  (LAYER METAL4 35 (WIDTH 0.2800) )
  (LAYER METAL4 36 (WIDTH 0.8000) )
  (LAYER METAL5 57 (WIDTH 1.6400) )
  (LAYER METAL5 58 (WIDTH 0.7400) )
  (LAYER METAL6 60 (WIDTH 1.0) )
)
(GATE_DEF
  (GATE USB20
    (RECT 0 0 100 75)
    (RECT 50 75 100 125)
    (TYPE digital)
    (OBSTRUCTION 4)
  )
)
(VIA_DEF
  (VIA VIA12 10
    (RECT 1 1 2 2)
    (RES 1.1)
    (CAP 0.2)
  )
)
```

Figure 5-5 Design-specific components of a design.

els called light_wl and heavy_wl are described. In the light_wl wireload model, the definition shows that for a net with a pincount of 3, the resistance of the net can be approximated to 1.5 and the capacitance can be approximated to 0.4; the units for resistance and capacitance are described in the header section.

In the layer definition section, the description of each layer is provided. This includes its width, orientation, resistance and capacitance per unit length and routing percentage allowed on the layer, if any. The layer definition also defines the hash id number associated with each metal layer; this hash id is later used in the PDEF file to reference the layer. Additional attributes such as WIDTH of the metal route can also be specified. In our example, one such layer defined has the layer name METAL1 with a WIDTH attribute of 0.23, and a hash id of 1. Other attributes that can be specified for a layer are:

- (**ORIENT** "0" | "90") : Specifies the preferred routing direction of the layer.
- (**RES** *res_per_unit_length*) : Specifies the resistance per unit length.
- (**CAP** *cap_per_unit_length*) : Specifies the capacitance per unit length.
- (**GAP** *edge_spacing*) : Specifies minimum edge spacing between routes on this layer.

Another design-specific component is the gate definition. A gate definition defines the shape of the abstract, whether rectangular or non-rectangular, its pins (one pin can exist on multiple physical layers), and the type of pin, such as iopad, custom, gnd, pwr or analog. The definition specifies the gate identifier and its attributes. Gate attributes include:

- (**RECT** *num num num num*) : Specifies the lower-left and upper-right coordinates of the gate.
- (**TYPE** iopad | custom | gnd | pwr | analog | digital)
- (**OBSTRUCTION** *layer_hash_id num num num num*) : Routing obstacle with the specified coordinates on the specified layer.
- (**RESTRICTION** fixed_logic | fixed_placement)

In the example shown in Figure 5-5, we have defined a gate with name USB20 and it has a non-rectangular geometry specified using two

RECT constructs. The gate type is `digital` and it has a routing obstruction on layer with a hash id of 4.

The via definition section defines the via names and their hash id numbers; these hash id numbers are used later in the PDEF file. In the PDEF example shown in Figure 5-5, a via `VIA12` is defined with a hash id of 10, via resistance of 1.1 and with a via capacitance of 0.2. It is rectangular in shape with the coordinates of the lower-left and upper-right specified as (1 1 2 2).

The last section in a PDEF file is the main part of the PDEF and it describes the physical and grouping information of the design in the form of clusters. A PDEF file can contain one or more clusters. Each cluster may define additional subclusters. For each cluster, it defines the cells that are part of the cluster and nets and buses that are part of the global routes.

An example of a cluster is shown in Figure 5-6. There is one cluster defined with the physical name `DR0012`. Its size is specified by the `RECT` attribute. For each cluster, its set of pins are defined using the `PIN` construct.

A cluster may have additional optional attributes that define its location, shape and soft macros (fences) if any.

- (**RECT** *num num num num*) : Defines rectangle location (lower-left and upper-right coordinates). Multiple RECT statements can be used to describe non-rectangular shapes.
- (**X_BOUNDS** *num num*) : Specifies the lower and upper bounds on the X axis where the cluster is placed.
- (**Y_BOUNDS** *num num*) : Specifies the lower and upper bounds on the Y axis where the cluster is placed.
- (**WIRELOAD** *name*) : Name of wireload model that applies to this cluster.
- (**MAX_HALF_PERIMETER** *num*) : Defines the maximum half sum of the X and Y perimeter of the cluster.
- (**MAX_X_PERIMETER** *num*) : Indicates the maximum X half perimeter of the soft macro boundng box.

```
(CLUSTER DR0012
  (RECT 0 0 2340 2340)
  (ORIGIN 0 0)
  (CONTENT_LOCATIONS relative)
  (PIN nreset)
  (PIN clk_32khz)
  (PIN x2_32khz)
  (PIN ntst)
  (PIN timer_ext)
  (PIN pwr_on)
  (PIN VDD
    (RECT 0 30 0.5600 40)
    (RECT 0 580 0.5600 590)
    (LAYER 42)
  )
  (SITE row_0 tsm3site "287 569.24" "H" "0" "0.66" "2675")
  (SITE row_1 tsm3site "287 564.2" "H" "180-mirror" "0.66" "2675")
  (SITE row_2 tsm3site "287 287" "H" "0" "0.66" "2675")
  (CELL top_core/qt32s1/qt32c1/BW1_INV39905_6
    (GATE_NAME INVX1)
    (ORIENT "180-mirror")
    (LOC 1461.1400 1239.5600)
  )
  (CELL top_core/qt32s1/qt32c1/BW1_INV_H24433
    (GATE_NAME INVX1)
    (ORIENT "0-mirror")
    (LOC 1098.1400 1365.5600)
  )
  (CELL top_core/vid1/vidfifo1/U973_C3_21
    (GATE_NAME AOI22X1)
    (ORIENT "180-mirror")
    (LOC 744.3800 1249.6400)
  )
  (CELL top_core/vid1/vidfifo1/U942_C1_7
    (GATE_NAME NAND4BX2)
    (ORIENT "0")
    (LOC 999.8000 1224.4400)
  )
  (CELL top_core/bufc1/tags_reg_6_2
    (GATE_NAME DFFRHQX1)
    (ORIENT "180")
    (LOC 1403.0600 1098.4400)
  )
```

Figure 5-6   A cluster definition.

- (**MAX_Y_PERIMETER** *num*) : Indicates the maximum Y half perimeter of the soft macro bounding box. The above three attributes are used to define soft macros (fences).

- (**RESTRICTION** *id*) : Specifies the type of restriction for the cluster. Legal values are -
  - FIXED_LOGIC : No logic can be changed or added to the cluster either by optimization or manually.
  - FIXED_PLACEMENT : Placement of cells is fixed; this is typical of hard macros (preplaced and prerouted macros).
  - FIXED_BOUNDS : The location of the cluster is fixed and cannot be moved.
  - FIXED_ROUTE : Routes cannot be changed.
  - EDIT_ROUTE : Routes can only be edited manually by user, not by a tool.
  - NO_NEW_ROUTE : No new routes can cross the boundary.
  - NO_NEW_CELLS : No new cells can be added to the cluster.
  - NO_NEW_BUFFERS : No new buffers can be added to the cluster.

- (**ROW_ORIENT** vertical | horizontal) : Specifies how the placement rows are oriented.

- (**CELL_ORIENT** "0" | "90" | "180" | "270" | "0-mirror" | "90-mirror" | "180-mirror" | "270-mirror") : Specifies the default orientation of all cells in the cluster. This orientation can be overriden on a cell by cell basis.

- (**OBSTRUCTION** *layer_number* [*num num num num*]) : Specifies the layer on which obstruction appears. If no shape is specified, the obstruction is on the entire cluster; otherwise the obstruction is of the shape specified.

- (**ORIGIN** *num num*) : Specifies the origin of the cluster with respect to the parent cluster.

- (**CONTENT_LOCATIONS** absolute | relative) : Specifies whether cell locations are relative to the cluster origin or absolute to to the design origin.

- (**CONTENT_ORDER** vertical | horizontal [*num*]) : Specifies the order in which cells in the cluster are to be placed. A wraparound limit may also be specified.

- (**CONNECTIVITY** complete | partial | unknown) : Specifies whether routes in cluster describe the entire connectivity or not.

- (**POWER** *name*) : Name of power domain to which cluster belongs.

There are seven pins defined in the cluster DR0012: nreset, clk_32khz, etc. There are additional optional attributes that can be specfied for a pin. These include:

- (**RECT** *num num num num*) : Specifies the geometry of the pin. Multiple RECT definitions can be used for specifying any non-rectangular pin geometry.

- (**LOC** *num num*) : This is the location of the pin with respect to the origin of the cluster.

- (**LAYER** *num*) : The hash id of the layer on which the pin belongs.

- (**SIDE** top | bottom | left | right) : The side on which the pin is on.

- (**ROW** *pos_integer*) : Constrains the pin to be within the specified row.

- (**COL** *num*) : Constrains the pin to be within the specified column.

- (**H_POS** *num*) : Specifies the ordering of pins in the horizontal direction.

- (**V_POS** *num*) : Specifies the ordering of pins in the vertical direction.

- (**TYPE** control | data | tri | gnd | pwr | clock | set | reset | test_en | scan_in | scan_out) : Specifies the type of pin.

An electrically equivalent pin may appear on multiple physical locations or layers; in such a case, the pin definition is repeated. Pin VDD shows the usage of attributes RECT and LAYER.

The PIN section is followed by the CELL section in which all cells that belong to the cluster are listed. For each cell, its instance name, its type, orientation and location are specified. For example, the cluster DR0012 contains the cell instance top_core/vid1/vidfifo1/ U973_C3_21 which is of type AOI22X1 and its orientation is "180-mirror" (rotation + reflection on Y axis) and its location is at (744.3800 1249.6400) with respect to the origin of the cluster. Other legal values of orientation are "0", "90", "180", "270", "0-mirror", "90-mirror" and "270-mirror".

A cell may optionally have additional attributes. These are:

- (**X_BOUNDS** *num num*) : Specifies the lower and upper bounds on the X axis indicating the bounds within which the cell is to be placed.
- (**Y_BOUNDS** *num num*) : Specifies the lower and upper bounds on the Y axis indicating the bounds within which the cell is to be placed.
- (**POWER** *string*) : Specifies the name of the power domain.
- (**RESTRICTION** fixed_logic | fixed_placement | fixed_bounds) : Specifies the restriction on the cell.

If a CELL definition does not appear within a cluster, then it is assumed that it belongs to the top level. An example of this is an IO cell.

The CELL definitions are followed by the NET constructs. A NET construct defines the net name, its attributes and any route information for the net. Some of the predefined attributes are:

- (**ROW** *num*) : Defines row within which the net is constrained to.
- (**COL** *num*) : Defines column within which the net is constrained to.
- (**TYPE** control | data | tri | gnd | pwr | clock | set | reset | test_en | analog | low_noise | hi_perf | reserved) : Can have multiple TYPE definitions and are all cumulative.
- (**PRI** *weighting_number*) : This is the routing priority of the net; 0 is the highest priority.
- (**LAYER** *number*) : The hash id of the layer that the net is constrained to. There can be multiple layer definitions, all of which are cumulative.
- (**RESTRICTION** fixed_logic | fixed_placement | no_new_buffers) : Restriction fixed_logic specifies that logic connected to the net cannot be changed in any way. Restriction fixed_placement specifies that placement of cells connected to the net cannot be changed in any way. Restriction no_new_buffers specifies that no buffers can be inserted to break up the net.

A ROUTE definition of a net provides details of how the net is routed. A route is broken up into segments. A route may optionally have the attribute:

- (**RESTRICTION** fixed | edit): Restriction value of fixed implies that no changes are allowed; edit implies that only user edits are allowed.

A segment is described using:

    (*pos_integer* <point> <point>)

where the pos_integer specifies the routing layer hash id and the points specify the coordinates of the end point of the segment. A point value of * implies the previous value.

Figure 5-7 shows the NET description part of the cluster definition. PNETS can also be present in a PDEF file. A pnet is a physical net that appears within a cluster. It has no corresponding equivalent in the logical netlist, and thus has a physical name as opposed to a net name. Pnets include special nets, such as power nets, that do not appear in the design netlist.

A PDEF file can also define buses; this is defined using a BUS construct. A bus defines a collection of nets or pins.

    (**BUS** *name* {*net_names*} | {*pin_names*})

Additional attributes that a BUS construct may have are:

- (**TYPE** control | data | tri) : Defines the specific kind of bus.
- (**SIDE** top | bottom | left | right) : Specifies which side the pins are on.
- (**LAYER** *hash_id_of_layer*) : Specifies layers that routes for nets or pins are constrained to use. There can be multiple LAYER definitions.

Another example of a PDEF file is shown in Figure 5-8, that for the clusters shown in Figure 5-2.

```
(NET top_core/xmem_done
  (ROUTE
     (1 (1681.4150 814.2400) (1682.2400 *))
     (* (1576.9700 820.4000) (1577.1350 *))
     (19 (1576.9700 *) (* 822.0800))
     (* (1668.7100 814.2400) (* 822.0800))
     (* (1682.2400 814.0750) (* 814.5200))
     (* (1689.1700 822.6400) (* 823.2000))
     (20 (1704.9025 826.8075) (* 828.6925))
     (19 (1704.7950 828.7500) (* 828.8500))
     (35 (1689.1700 814.2400) (* 822.6400))
  )
)
(NET top_core/qt32s1/qt32c1/new_bit_cnt[3]
  (ROUTE
     (1 (1116.2900 1411.7600) (1118.7650 *))
     (* (1115.6300 1383.9000) (* 1384.7400))
     (* (1221.0650 1090.8800) (* 1097.0400))
  )
)
))
```

Figure 5-7   Nets description in a cluster.

```
(CLUSTERFILE
  (PDEFVERSION "IEEE 1481-1999")
  (DESIGN "FIGURE5dot2")
  (DATE "February 17, 2005 09:04:56")
  (VENDOR "Star Galaxy Automation, Inc")
  (PROGRAM "StarLayTool")
  (VERSION "2005.01.01")
  (DIVIDER /)
  (PIN_DELIMITER :)
  (BUS_DELIMITER [ ] )
  (NETLIST_TYPE VHDL)
  (DESIGN_FLOW gates)
  (DESIGN_FLOW cluster_locations)
  (RESISTANCE_UNIT 0.1)
  (CAPACITANCE_UNIT 0.1)
  (DISTANCE_UNIT 0.1)
  (LAYER_DEF
    (LAYER Met1 10 (RES 0.01) (CAP 0.011) (WIDTH 0.21))
    (LAYER Met2 20 (RES 0.011) (CAP 0.015) (WIDTH 0.21))
    (LAYER Met3 30 (RES 0.015) (CAP 0.018) (WIDTH 0.22))
  )
  (CLUSTER TOP
    (RECT 0 0 900 700) (ORIGIN 0 0)
    (CONNECTIVITY partial)
    (PIN DATAIN) (PIN Q) (PIN CLOCK (LAYER 40) )
    (CLUSTER C1 (RECT 0 250 500 700) (RESTRICTION fixed_logic))
    (CLUSTER C2
      (RECT 0 0 500 250) (X_BOUNDS 0 800 ) (Y_BOUNDS 0 300)
      (CELL U1 (GATE_NAME BUFX20) (ORIENT "0-mirror") (LOC 20 20))
      (CELL U2 (GATE_NAME SDFFX1) (ORIENT "180") (LOC 80.5 72.1)
        (RESTRICTION fixed_placement)
      )
    )
    (CLUSTER C3 (RECT 500 0 900 700)
      (CLUSTER C4 (RECT 510 20 890 150)
        (NET N1 (TYPE data) (PRI 5) (RESTRICTION fixed_placement)
          (ROUTE
            (1 (10 12) (12 *))
            (* (26 *) (* 52))
            (RESTRICTION fixed)
          )
        )
      )
    )
  )
)
```

Figure 5-8   Another PDEF example.

5.3   Complete Syntax

This section describes the complete syntax[1] of a PDEF file.

Comments come in two forms: `//` starts a comment until end of line, while `/* . . . */` is a multi-line comment.

In the following syntax, bold characters such as **(** and **[** are part of the syntax. All constructs are arranged alphabetically and the start symbol is `PDEF_file`.

```
alpha ::= upper | lower

attribute ::=
   (non_numeric_identifier)
   | (identifier identifier {identifier})
   | (identifier qstring {qstring})
   | (identifier pos_number {pos_number})
   | (identifier number {number})
   | (identifier pos_integer {pos_integer})
   | (identifier pos_integer qstring {qstring})
   | (identifier pos_integer identifier {identifier})
   | (identifier pos_integer path {path})
   | (identifier pos_integer pos_number {pos_number})
   | (identifier pos_integer number {number})
   | (identifier identifier qstring {qstring})
   | (identifier identifier pos_integer {pos_integer})
   | (identifier identifier identifier qstring)

bit_identifier ::=
   identifier
   | <identifier><prefix_bus_delim><digit>
     {<digit>}[<suffix_bus_delim>]

bus ::= nbus | pbus

bus_delimiter_def ::=
   (BUS_DELIMITER prefix_bus_delim [ suffix_bus_delim ])
```

---

1. Syntax is reprinted here with permission from IEEE Std. 1481-1999, Copyright 1999, by IEEE. All rights reserved.

```
cell ::= (CELL inst_name {attribute})

cluster ::=
  (CLUSTER physical_name {attribute} {gap} {cluster_object})

cluster_object ::=
  pin | node | cell | spare_cell | skip | net | pnet
  | bus | cluster

cont_number ::= * | number

cont_pos_integer ::= * | pos_integer

date ::= (DATE qstring)

decimal ::= [sign]<digit>{<digit>}.{<digit>}

designator ::= pos_integer | all

design_flow ::= (DESIGN_FLOW identifier | qstring)

design_name ::= (DESIGN name)

digit ::= 0 - 9

escaped_char ::= \<escaped_char_set>

escaped_char_set ::= special_char | "

exp ::= <radix><exp_char><integer>

exp_char ::= E | e

float ::= decimal | fraction | exp

fraction ::= [sign].<digit>{<digit>}

gap ::= (GAP number gap_constraint {attribute})

gap_constraint ::= (row_col_designator)

gate ::= (GATE identifier {attribute} {gate_pin})
```

gate_def ::= **(GATE_DEF** {gate}**)**

gate_pin ::= **(PIN** bit_identifier {attribute}**)**

hchar ::= . | **/** | : | |

header_def ::=
  pdef_version
  design_name
  date
  vendor
  program_name
  program_version
  hierarchy_divider_def
  pin_limiter_def
  bus_delimiter_def
  netlist_type
  {design_flow}
  {file_attribute}

hierarchy_divider_def ::= **(DIVIDER** hier_delim**)**

hier_delim ::= hchar

identifier ::= <identifier_char>{<identifier_char>}

identifier_char ::= escaped_char | alpha | digit | _

inst_name ::= logical_name

integer ::= [sign]<digit>{<digit>}

layer ::= **(LAYER** name pos_integer {attribute}**)**

layer_def ::= **(LAYER_DEF** {layer}**)**

logical_name ::= path | prefixed_path

lower ::= **a - z**

name ::= qstring | identifier

nbus ::= **(BUS** physical_name {attribute} {net_item}**)**

```
nbus_ref ::= physical_ref

neg_sign ::= -

net ::= (NET net_name {attribute} [route])

netlist_type ::= (NETLIST_TYPE identifier)

net_item ::= net | pnet | (BUS nbus_ref)

net_name ::= logical_name

node ::= (NODE bit_identifier {attribute})

node_ref ::= <physical_ref><pin_delim><bit_identifier>

non_numeric_char ::= escaped_char | alpha | _

non_numeric_identifier ::=
  {<identifier_char>}<non_numeric_char>{<identifier_char>}

number ::= integer | float

path ::=
  [hier_delim]<bit_identifier>{<partial_path>}{hier_delim}

partial_path ::= <hier_delim><bit_identifier>

partial_physical_ref ::= <hier_delim><physical_name>

pbus ::= (BUS physical_name {attribute} {pin_item})

pbus_ref ::= physical_ref

PDEF_file ::=
  (CLUSTERFILE
  header_def
  [wire_load_def]
  [layer_def]
  [via_def]
  [gate_def]
  {cluster})
```

```
pdef_version ::= (PDEFVERSION qstring)

physical_name ::= name

physical_ref ::=
  .<partial_physical_ref>{<partial_physical_ref>}
  | <physical_name>{<partial_physical_ref>}

pin ::= (PIN pin_name {attribute})

pincount ::= (PINCOUNT pos_integer {attribute})

pin_delim ::= hchar

pin_delimiter_def ::= (PIN_DELIMITER pin_delim)

pin_item ::= pin | node | (BUS pbus_ref)

pin_name ::= <inst_name><pin_delim><bit_identifier>

pnet ::= (PNET physical_name {attribute} [route])

point ::= route_pin | (cont_number cont_number)

pos_decimal ::= <digit>{<digit>}.{<digit>}

pos_exp ::= <pos_radix><exp_char><integer>

pos_float ::= pos_decimal | pos_fraction | pos_exp

pos_fraction ::= .<digit>{<digit>}

pos_integer ::= <digit>{<digit>}

pos_number ::= pos_integer | pos_float

pos_radix ::= pos_integer | pos_decimal | pos_fraction

pos_sign ::= +

prefixed_path ::= prefix_index path

prefix_bus_delim ::= [ | { | ( | < | : | .
```

```
prefix_index ::= pos_integer

program_name ::= (PROGRAM qstring)

program_version ::= (VERSION qstring)

qstring ::= "{<qstring_char>}"

qstring_char ::= special_char | alpha | digit | white_space | _

radix ::= integer | decimal | fraction

route ::= (ROUTE {attribute} {route_item})

route_item ::= {segment} | {via_ref}

route_pin ::= (route_pin_ref [(cont_number cont_number)])

route_pin_ref ::= PIN pin_name | NODE node_ref

row_col ::= ROW | COL

segment ::= (cont_pos_integer point point)

sign ::= pos_sign | neg_sign

skip ::= (SKIP pos_number {attribute})

spare_cell ::= (SPARE_CELL physical_name {attribute})

special_char ::= ! | # | $ | % | & | ' | ( | ) | * | + | , | - | .
    | / | : | ; | < | = | > | ? | @ | [ | \ | ] | ^ | ' | { | | | } | ~

suffix_bus_delim ::= ] | } | ) | >

upper ::= A - Z

vendor ::= (VENDOR qstring)

via ::= (VIA name pos_integer {attribute})

via_def ::= (VIA_DEF {via})
```

via_ref ::= **(VIA** pos_integer point**)**

wire_load ::= **(WIRE_LOAD** name {attribute} {pincount}**)**

wire_load_def ::= **(WIRE_LOAD_DEF** {wire_load}**)**

white_space ::= space | tab

❑

VCD

T his chapter describes the Value Change Dump format (VCD). It is part of the IEEE Std 1364.

6.1 Basics

A VCD file contains ASCII descriptions of waveforms in a design. That is, it lists the variables and the value changes with their time stamps. It is mainly used as an exchange medium to pass waveforms of a design between multiple tools. For example, a VCD file may be produced by a formal verification tool and used by a power analysis tool to compute power used in a design. But quite often, a VCD file is produced during simulation.

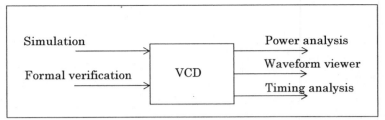

Figure 6-1   VCD is a tool exchange medium.

There are two types of VCD files:

i.   *Four state VCD*: Shows the four values 0, 1, **x**, **z** with no strength information.

ii. *Extended VCD*: Shows value changes with strength information.

A VCD file is free format and white space is used to separate commands.

6.2   Four State VCD Example

The format of a four state VCD file is as follows:

```
Date and time
Version
Timescale
Scope of module
Variable definitions
End of header
Simulation time
Values
Simulation time
Values
Simulation time
Values
   . . .
```

```
$date                                    $dumpvars
Fri Sep 27 16:23:58 1996                 1#
$end                                     0$
$version                                 b1 !
Verilog HDL Simulator 1.0                b10 "
$end                                     b101 +
$timescale                               1(
100ps                                    0'
$end                                     1&
$scope module Test $end                  1)
$var parameter 32 ! ON_DELAY             0*
$end                                     $end
$var parameter 32 " OFF_DELAY            #10
$end                                     0#
$var reg 1 # Clock $end                  0)
$var reg 1 $ UpDn $end                   #30
$var wire 1 % Cnt_Out [0] $end           1#
$var wire 1 & Cnt_Out [1] $end           1)
$var wire 1 ' Cnt_Out [2] $end           b100 +
$var wire 1 ( Cnt_Out [3] $end           b101 +
$scope module C1 $end                    #40
$var wire 1 ) Clk $end                   0#
$var wire 1 * Up_Down $end               0)
$var reg 4 + Count [0:3] $end            #60
$var wire 1 ) Clk $end                   1#
$var wire 1 * Up_Down $end               1)
$upscope $end                            b100 +
$upscope $end                            b101 +
$enddefinitions $end                     #70
#0                                       0#
(continued next column)                  . . .
```

Figure 6-2 A four state VCD example.

Figure 6-2 shows an example of a four state VCD file. The lines:

```
$date
Fri Sep 27 16:23:58 1996
$end
```

specify the date and the time at which the file was generated. The lines:

```
$version
Verilog HDL Simulator 1.0
$end
```

specify the tool and version number that generated the VCD file. The lines:

```
$timescale
100ps
$end
```

specify the timescale of the time values that appear in the file. The lines:

```
$scope module Test $end
```

specifies the scope of the variables which are listed next. The scope can be a top-level module (**module**), module instance (**module**), task (**task**), function (**function**), a named sequential block (**begin**) or a named parallel block (**fork**). The line:

```
$upscope $end
```

changes to the next higher level of hierarchy. Variables whose values are dumped are declared following its scope declaration. Each variable is associated with an identifier code, which is a symbol from the printable ASCII character set. The identifier code could be one or more characters. The variable declaration includes its type (reg or wire) and its size as well.

```
$var wire 1 * Up_Down $end
$var reg 4 + Count [0:3] $end
```

Variable Up_Down is declared with an id code of * and it is also declared to be a wire of size 1 bit. Variable Count is declared with an identifier code of + and it is a reg variable with a size of 4 bits. The identifier codes are used later instead of the variable names. So a reference to * refers to the variable Up_Down, etc.

```
$enddefinitions $end
```

marks the end of the header and definitions section. What follows next is the main core of the VCD file.

First, a simulation time is specified:

```
#0
```

which is followed by variables (using their identifier code) and their values at that simulation time. This set of variables ends when the next simulation time is specified. The simulation time specified is the absolute value of the time at which the variables have the specified values. Thus the value changes are of the format:

```
#simulation_time
value identifier_code
value identifier_code
. . .
#simulation_time
value identifier_code
value identifier_code
. . .
#simulation_time
value identifier_code
value identifier_code
. . .
$end
```

So the lines in the VCD file:

```
1#
0$
b1 !
b10 "
b101 +
```

represent variable values. The variable with identifier code # has value 1, the variable with identifier code $ has value 0, the variable with identifier code ! has 1, the variable with identifier code " has value 10, and so on. Values for all variables are specified using the four values 0, 1, x, z, except for real variables for which real values are used.

For scalar variables, there is no space between the value and its identifier code. For vector variables, there is a one space character between the value and its identifier code.

Note that only value changes are logged in a VCD file. Data in a VCD file is case-sensitive.

A comment section can also appear in a VCD file. Here is an example.

```
$comment A comment can be on a
single line or span multiple lines
$end
```

There are four additional simulation keywords that may appear as part of the value change list. These are:

- $dumpall: Specifies the current value of all variables being dumped.

```
$dumpall
1*
0!
x(
$end
```

- $dumpoff: Indicates all variables that are dumped with x values.

```
$dumpoff
x*
x+
x}
$end
```

- $dumpon: Indicates resumption of dumping and lists the current value of all variables.

```
$dumpon
b10 %
B1011 &
1#
$end
```

- $dumpvars: Lists the initial value of variables being dumped.

```
$dumpvars
x5
0 (
$end
```

6.3   Four State VCD Syntax

Listed alphabetically here is the complete syntax[1] of a VCD file. The start symbol is *value_change_dump_definitions*.

```
declaration_command ::=
    vcd_declaration_comment
  | vcd_declaration_date
  | vcd_declaration_enddefinitions
  | vcd_declaration_scope
  | vcd_declaration_timescale
  | vcd_declaration_upscope
  | vcd_declaration_version
  | vcd_declaration_vars

identifier_code ::=
  { ASCII character }

index ::=
  decimal_number

reference ::=
    identifier
  | identifier [ bit_select_index ]
  | identifier [ msb_index : lsb_index ]

scalar_value_change ::=
  value identifier_code

scope_type ::=
    begin
  | fork
```

---

1. Syntax reprinted with permission from IEEE Std. 1364-2001, Copyright 2001, by IEEE. All rights reserved.

```
                    | function
                    | module
                    | task

        simulation_command ::=
            vcd_simulation_dumpall
          | vcd_simulation_dumpoff
          | vcd_simulation_dumpon
          | vcd_simulation_dumpvars
          | simulation_time
          | value_change

        simulation_time ::=
          # decimal_number

        size ::=
          decimal_number

        time_number ::=
          1 | 10 | 100

        time_unit ::=
          s | ms | us | ns | ps | fs

        value ::=
          0 | 1 | x | X | z | Z

        value_change ::=
            scalar_value_change
          | vector_value_change

        value_change_dump_definitions ::=
          { declaration_command } { simulation_command }

        var_type ::=
            event | integer | parameter | real | reg | supply0 | supply1
          | time | tri | triand | trior | trireg | tri0 | tri1 | wand
          | wire | wor

        vcd_declaration_comment ::=
          $comment comment_text $end
```

```
vcd_declaration_date ::=
  $date date_text $end

vcd_declaration_enddefinitions ::=
  $enddefinitions $end

vcd_declaration_scope ::=
  $scope scope_type scope_identifier $end

vcd_declaration_timescale ::=
  $timescale time_number time_unit $end

vcd_declaration_upscope ::=
  $upscope $end

vcd_declaration_vars ::=
  $var var_type size identifier_code reference $end

vcd_declaration_version ::=
  $version version_text system_task $end

vcd_simulation_dumpall ::=
  $dumpall { value_changes } $end

vcd_simulation_dumpoff ::=
  $dumpoff { value_changes } $end

vcd_simulation_dumpon ::=
  $dumpon { value_changes } $end

vcd_simulation_dumpvars ::=
  $dumpvars { value_changes } $end

vector_value_change ::=
    b binary_number identifier_code
  | B binary_number identifier_code
  | r real_number identifier_code
  | R real_number identifier_code
```

6.4   Extended VCD Example

The format of an extended VCD file, which is very similar to a four state VCD file, is as follows:

```
Date and time
Version
Timescale
Scope of module
Variable definitions
End of header
Simulation time
Values
Simulation time
Values
Simulation time
Values
. . .
Final simulation time
```

Figure 6-3 shows an example of an extended VCD file. The lines:

```
$comment
File created using the following command:
vcd file count.evcd -dumpports
$end
```

show the comment section; the comments describe the command that was used to created the extended VCD file. The following lines show the timestamp when the file was created:

```
$date
Thu Mar  4 08:44:00 2004
$end
```

The $version keyword is used to specify the version of the tool used to generate the extended VCD file.

```
$version
dumpports VerilogSim Version 1.0a
$end
```

```
$comment                                    #12
File created using the following            pU 0 6 <0
command:                                    #13
vcd file count.evcd -dumpports              pD 6 0 <0
$end                                        #15
$date                                       pU 0 6 <0
Thu Mar  4 08:44:00 2004                    . . .
$end                                        #49
$version                                    pD 6 0 <0
dumpports VerilogSim Version 1.0a           #50
$end                                        $dumpportsall
$timescale                                  pH 0 6 <4
1ns                                         pL 6 0 <3
$end                                        pH 0 6 <2
$scope module test_counter $end             pL 6 0 <1
$scope module ucud $end                     pD 6 0 <0
$var port 1 <0 clk $end                     pD 6 0 <5
$var port 1 <1 count [0] $end               $end
$var port 1 <2 count [1] $end               pU 0 6 <5
$var port 1 <3 count [2] $end               #51
$var port 1 <4 count [3] $end               pU 0 6 <0
$var port 1 <5 up_down $end                 pL 6 0 <4
$upscope $end                               pH 0 6 <3
$upscope $end                               #52
$enddefinitions $end                        pD 6 0 <0
#10                                         #54
$dumpports                                  pU 0 6 <0
pH 0 6 <4                                   pH 0 6 <4
pL 6 0 <3                                   #55
pH 0 6 <2                                   pD 6 0 <0
pL 6 0 <1                                   #57
pD 6 0 <0                                   pU 0 6 <0
pD 6 0 <5                                   pL 6 0 <4
$end                                        pL 6 0 <3
(continued next column)                     pL 6 0 <2
                                            pH 0 6 <1
                                            . . .
                                            $vcdclose #151 $end
```

Figure 6-3   An extended VCD file.

The following directive specifies the timescale of time values that appear in the extended VCD file.

```
$timescale
1ns
$end
```

The lines:

```
$scope module test_counter $end
$scope module ucud $end
```

define the scope of the variables being dumped.

The *var declaration* defines the variable being dumped. It specifies that the variable is a port (can only be a port), the number of bits in the port and an identifier code (the number following the < character).

```
$var port 1 <0 clk $end
$var port 1 <1 count [0] $end
$var port 1 <2 count [1] $end
$var port 1 <3 count [2] $end
$var port 1 <4 count [3] $end
$var port 1 <5 up_down $end
$var port [3:0] <6 gray_code $end
```

A bus can also be explicitly specified as shown by the declaration of `gray_code`. The left index is the most significant bit and the right index is the least significant bit.

The following directive changes the scope to the next higher level in design hierarchy.

```
$upscope $end
```

The line:

```
$enddefinitions $end
```

specifies the end of the header and definitions section. What follows next is the sequence of simulation time followed by a list of values, which is repeated over and over. The line:

```
#50
```

specifies the simulation time for which the value changes are listed next. The units are as specified with the $timescale directive. The value changes specify the logic value and any strength information. In the lines:

```
pH06 <4
pL60 <3
pH06 <2
pL60 <1
```

the first character p specifies that it is a port. This is immediately followed (with no space) by the state character. A state character is specified based on whether the port is an input port, output port or has an unknown direction. For an input port, the state character is:

- D: low
- U: high
- N: unknown
- Z: three-state
- d: low with two or more drivers active
- u: high with two or more drivers active

For an output port, the state character is:

- L: low
- H: high
- X: unknown or don't care
- T: three-state
- l: low with two or more active drivers
- h: high with two or more active drivers

For an unknown direction, the state character is:

- 0: low (both input and output are active with 0 value)
- 1: high (both input and output are active with 1 value)
- ?: unknown
- F: three-state (input and output unconnected)
- A: unknown (input 0 and output 1)

- **a**: unknown (input 0 and output **x**)
- **B**: unknown (input 1 and output 0)
- **b**: unknown (input 1 and output **x**)
- **C**: unknown (input **x** and output 0)
- **c**: unknown (input **x** and output 1)
- **f**: unknown (input and output three-stated)

The state character is followed by a 0-value strength specification and a 1-value strength specification and then followed by the identifier code. A strength specification has the following values:

- **0**: highz
- **1**: small
- **2**: medium
- **3**: weak
- **4**: large
- **5**: pull
- **6**: strong
- **7**: supply

So in the value specification,

> p**H** 0 6 <4

it specifies a value for port 4 (which is count[3]) which has an high value with a highz 0-value strength and a strong 1-value strength.

The $vcdclose construct, which appears at the end of file, specifies the final simulation time at which the extended VCD file is closed.

> $**vcdclose** #151 $**end**

The file was closed at time 151ns.

6.5   Extended VCD Syntax

Described here is the complete syntax[1] of the extended VCD file. The syntax has been arranged alphabetically. The starting term is `value_change_dump_definitions`. Terminal symbols appear in *italics*.

```
close_text ::=
  final_simulation_time

command_text ::=
    comment_text | close_text | date_section | scope_section
  | timescale_section | var_section | version_section

comment_text ::=
  {ASCII_character}

date_section ::=
  date_text

date_text ::=
  day month date time year

declaration_command ::=
  declaration_keyword [ command_text ] $end

declaration_keyword ::=
    $comment | $date | $enddefinitions | $scope | $timescale
  | $upscope | $var | $vcdclose | $version

dumpports_command ::=
  $dumpports ( scope_identifier, file_pathname )

file_pathname ::=
    string_literal
  | variable
  | expression
```

---

1. Syntax reprinted with permission from IEEE Std. 1364-2001, Copyright 2001, by IEEE. All rights reserved.

```
identifier ::=
  {printable_ASCII_character}

identifier_code ::=
  <{integer}

index ::=
  decimal_number

input_value ::=
  D | U | N | Z | d | u

number ::=
  1 | 10 | 100

output_value ::=
  L | H | X | T | l | h

port_value ::=
    input_value
  | output_value
  | unknown_direction_value

reference ::=
  port_identifier

scope_section ::=
  scope_type scope_identifier

scope_type ::=
  module

simulation_command ::=
    simulation_keyword { value_change } $end
  | $comment [ comment_text ] $end
  | simulation_time
  | value_change

simulation_keyword ::=
    $dumpports
  | $dumpportsoff
  | $dumpportson
  | $dumpportsall
```

```
simulation_time ::= #decimal_number

size ::=
  1 | vector_index

strength_component ::=
  0 | 1 | 2 | 3 | 4 | 5 | 6 | 7

timescale_section ::=
  number time_unit

time_unit ::=
  fs | ps | ns | us | ms | s

unknown_direction_value ::=
  0 | 1 | ? | F | A | a | B | b | C | c | f

value ::=
  pport_value 0_strength_component 1_strength_component

value_change ::=
  value identifier_code

value_change_dump_definitions ::=
  { declaration_command } { simulation_command }

var_section ::=
  var_type size identifier_code reference

var_type ::=
  port

vector_index ::=
  [ msb_index : lsb_index ]

version_section ::=
  version_text
```

```
version_text ::=
  version_identifier { dumpports_command }
```

❑

Bibliography

1. IEEE Std 1497-2001, *IEEE Standard for Standard Delay Format (SDF) for the Electronic Design Process*.

2. IEEE Std 1364-2001, *IEEE Standard Verilog Hardware Description Language*.

3. IEEE Std 1481-1999, *IEEE Standard for Integrated Circuit (IC) Delay and Power Calculation System*.

4. IEEE Std 1076.4-1995, *IEEE Standard for VITAL Application-Specific Integrated Circuit (ASIC) Modeling Specification*.

5. *LEF/DEF Langauge Reference*, Product Version 5.6, Sept 2004, Cadence Design Systems.

6. IEEE Std 1076-2002, *IEEE Standard VHDL Language Reference Manual*.

❑

Index

❑